THE PHILOSOPHY OF THE WEATHER, AND A GUIDE TO ITS CHANGES

LECTOR HOUSE PUBLIC DOMAIN WORKS

THE PHILOSOPHY OF THE WEATHER, AND A GUIDE TO ITS CHANGES

THOMAS BELDEN BUTLER

ISBN: 978-93-89701-20-3

Published: 1856

LECTOR HOUSE LLP
E-MAIL: lectorpublishing@gmail.com

THE PHILOSOPHY OF THE WEATHER.
AND A GUIDE TO ITS CHANGES.

BY

T. B. BUTLER.

1856.

INTRODUCTION.

The atmospheric conditions and phenomena which constitute "The Weather" are of surpassing interest. Now, we rejoice in the genial air and warm rains of spring, which clothe the earth with verdure; in the alternating heat and showers of summer, which insure the bountiful harvest; in the milder, ripening sunshine of autumn; or the mantle of snow and the invigorating air of a moderate winter's-day. Now, again, we suffer from drenching rains and, devastating floods, or excessive and debilitating heat and parching drought, or sudden and unseasonable frost, or extreme cold. And now, death and destruction come upon us or our property, at any season, in the gale, the hurricane, or the tornado; or a succession of sudden or peculiar changes blight our expected crops, and plant in our systems the seeds of epidemic disease and death. These, and other normal conditions, and varied changes, and violent extremes, potent for good or evil, are continually alternating above and around us. They affect our health and personal comfort, and, through those with whom we are connected, our social and domestic enjoyments. They influence our business prosperity directly, or indirectly, through our near or remote dependence upon others. They limit our pleasures and amusements—they control the realities of to-day, and the anticipations of to-morrow. None can prudently disregard them; few can withhold from them a constant attention. Scientific men, and others, devote to them daily hours of careful observation and registration. Devout Christians regard them as the special agencies of an over-ruling Providence. The prudent, fear their sudden, or silent and mysterious changes; the timid, their awful manifestations of power; and they are, to each and all of us, ever present objects of unfailing interest.

This *interest* finds constant expression in our intercourse with each other. A recent English writer has said: "The germ of meteorology is, as it were, innate in the mind of every Englishman—the weather is his first thought after every salutation." In the qualified sense in which this was probably intended, it is, doubtless, equally true of us. Indeed, it is often not only a "first thought" *after* a salutation, but a part of the salutation itself—an offspring of the same friendly feeling, or a part of the same habit, which dictates the salutation—an expression of sympathy in a subject of common and absorbing interest—a sorrowing or rejoicing with those who sorrow or rejoice in the frowns and smiles of an ever-changing, ever-influential atmosphere.

If consistent with our purpose, it would be exceedingly interesting to trace the varied forms of expression in use among different classes and callings, and see how indicative they are of character and employment.

The sailor deals mainly with the winds of the hour, and to him all the other phases of the weather are comparatively indifferent. He speaks of airs, and breezes, and squalls, and gales, and hurricanes; or of such appearances of the sky as prognosticate them. The citizens, whose lives are a succession of *days*, deal in such adjectives as characterize the weather of *the day*, according to their class, or temperament, or business; and it is pleasant, or fine, or *very* pleasant or fine; beautiful, delightful, splendid, or glorious; or unpleasant, rainy, stormy, dismal, dreadful or horrible. The farmer deals with the weather of considerable periods; with forward or backward *seasons*, with "cold snaps" or "hot spells," and "wet spells" or "dry spells." And there are many intermediate varieties. The acute observer will find much in them to instruct and amuse him, and will probably be surprised to find how much they have to do with his "first impressions" of others.

But I have a more important object in view. I propose to deal with "The Philosophy *of the Weather*" — to examine the nature and operation of the arrangements from which the phenomena result; to strip the subject, if possible, of some of the complication and mystery in which traditionary axioms and false theories continue to envelop it; to endeavor to grasp *its principles*, and unfold them in a plain, concise, and systematic manner, to the comprehension of "*the many*," who are equal partners with the scientific in its practical, if not in its philosophic interest; and to deduce a few general rules by which its changes may be understood, and, ultimately, to a considerable extent, foreseen.

This is not an easy, perhaps not a prudent undertaking. Nor is my position exactly that of a volunteer. A few words seem necessary, therefore, by way of apology and explanation.

In the fall of 1853, in the evening of a fair autumnal day, I started for Hartford, in the express train. Just above Meriden, an acquaintance sitting beside me, who had been felicitating himself on the prospect of fine weather for a journey to the north, called my attention to several small patches of scud — clouds he called them — to the eastward of us, between us and the full clear moon, which seemed to be enlarging and traveling south — and asked what they meant.

"Ah!" said I, "they are scud, forming over the central and northern portions of Connecticut, induced and attracted by the influence of a storm which is passing from the westward to the eastward, over the northern parts of New England, and are traveling toward it in a southerly surface wind, which we have run into. They seem to go south, because we are running north faster than they. You see them at the eastward because they are forming successively as the storm and its influence passes in that direction, and are most readily seen in the range of the moon; but when we reach Hartford you will see them in every direction, more numerous and dense, running north to underlie that storm."

I had seen such appearances too many times to be deceived. It was so. When we arrived at Hartford they were visible in all directions, running to the northward at the rate of twenty-five miles an hour. In the space of forty minutes we had passed from a clear, calm atmosphere (and which still remained so), into a cloudy, damp air, and brisk wind blowing in the same direction we were traveling, and toward a heavy storm. My friend passed on, and met the southern edge of the rain at Deerfield, and had a most unpleasant journey during the forenoon of the next day. Taking the cars soon afterwards, in the afternoon, for the south, I found him on his return.

"Shall I have fair weather now till I get home?" said he.

"There are no indications of a storm here, or at present," I replied, "but we may observe them elsewhere, and at nightfall."

He kept a sharp look-out, and, as we neared New Haven, discovered faint lines of cirrus cloud low down in the west, extending in parallel bars, contracting into threads, up from the western horizon, in an E. N. E. direction toward the zenith.

"Now, what is that?" said he.

"The eastern outlying edge of a N. E. storm, approaching from the W. S. W. It is now raining from 150 to 200 miles to the westward of the eastern extremity of those bars of cirrus-condensation; perhaps more, perhaps less; and under those bars of condensation the wind is attracted, and is blowing from the N. E. toward the body of the storm, and where the condensation is sufficiently dense to drop rain. That dense portion will reach here, and it will rain from twelve to fifteen hours hence. As we pass along the shore, and run under that out-lying advance cirrus-condensation, we shall see that the vessels in the Sound have the wind from the N. E., freshening, but we shall continue to have this light and scarcely-perceptible air from the northward for a time—*the N. E. wind always setting in toward an approaching storm, out on the Sound, much sooner than upon the land.*"

As we approached the storm, and the storm us, the evidence of denser condensation at the west, and of wind from the east, blowing toward it, became more apparent. The fore and aft vessels were running "up Sound" with "sheet out and boom off," before a fresh N. E. breeze, and my friend was astonished.

"I must understand this," said he; "how is it?"

"All very simple. The page of nature spread out above us is intelligible to him who will attentively study it. The laws which produce the impressions and changes upon that page, are few and comprehensible. Although there is great variety, even upon the limited portion which is bounded by our horizon, there is also substantial uniformity; and, although the changes are always extensive, often covering an area of one thousand miles or more, and our vision can not extend in any direction more than from thirty to fifty, yet those changes are always, to a considerable extent, intelligible, and may often be foreseen."

"Has meteorology made such progress?"

"By no means. It has, indeed, been raised to the dignity of a science, and professorships endowed for its advancement. Some books have been written, and many theories broached in relation to it; and innumerable observations of the states of the barometer and thermometer, of the clouds, and the quantity of fallen rain, and the direction and force of the wind—made and recorded simultaneously in different countries—have been published and compared; and a great many important facts established, and tables of '*means*' constructed, and just inferences drawn, yet the *few and simple arrangements* upon which all the phenomena depend, and *their philosophy*, have not yet been clearly elicited or understood."

"Have not the 'American Association for the Advancement of Science' arrived at some definite and sound conclusion upon the subject?"

"No; it has been with them, for many years, an interesting subject for papers and debate. Some very valuable articles, upon particular topics, or branches of the subject, have been read and published. But the *Cyclonologists*, as they term themselves, and who seem to think the great question is, '*Are storms whirlwinds?*' appear with new editions and phases of their favorite views as regularly as the annual meeting recurs; and, though they have not convinced, they seem to have silenced their opponents. The only conclusion, however, judging from their debates, to which the Association appear to have come with any considerable unanimity, is, that they are yet without sufficient *authentic observations* and well-established facts, to authorize the adoption of the Huttonian, Daltonian, Gyratory, or Aspiratory, or any of the other numerous theories which abound. And they are right. The subject is mystified by these theories and speculations of the study, founded on barometrical and thermometrical records, and the direction and force of the surface winds.

"The qualities of heat were among the earlier discoveries of science, and all the phenomena of the weather were forthwith attributed to its influence. Hastily-formed and erroneous views of its power, and the manner of its action in particular localities, and under particular circumstances, have retained the credence accorded to them when first announced, although subsequent discoveries have shown their fallacy; some new theory of *modification* having been invented to reconcile the discrepancies as soon as they appeared. Perhaps it is not too much to say (however it may seem to one not thoroughly acquainted with the subject, who does not know that the *primary* and secondary modifying hypotheses found in Kämtz, may be counted by hundreds) that there is not remaining in any other science, and possibly in all others, an equal amount of false and absurd theory, and of forced and unnatural grouping of admitted facts to sustain it, as in meteorology as at present taught and received. Astronomy, as a science, is almost perfected—the nature, and size, and orbits, of the distant worlds around us are known—while constant changes and alternating atmospheric conditions, which all occur *within less than six miles of us*, affecting all our important interests, and obvious to our senses, although much talked off, and made the objects of many theories, are but little understood."

"How, then, did you acquire the information you seem to possess?"

"By studying '*the countenance of the sky*,' for in no other way has such information ever been, or can it ever be, acquired. By a long-continued, daily, and sometimes hourly observation of the clouds and currents of the atmosphere, in connection with such reports of the then state of the weather elsewhere, as have fallen under my notice, and the effect of its changes upon the animal creation—for very much can be learned from them. Yonder flock of black ducks that sit on that inshore rock, above the tide—the wildest and most suspicious of all their tribe—although the air is calm about them, know well that a storm is at hand. They probably both see and feel it. As twilight approaches they will fly away inland, forty or fifty miles perhaps, and settle among the lilies or grass which surround some fresh-water pond, certain of remaining while the storm lasts, and for one day at least, out of danger, and undisturbed. Many a time, in my boyhood, have I heard, in the stillness of evening, the whistling of their wings, as they swept up the Connecticut valley, to seek, on the borders of the coves, and in the creeks of the meadows, a concealed and safe feeding-place during a coming storm. And many a time in the autumn, after they had all passed down for the season, when the indications of an approaching storm were clearly visible at nightfall, have I waited for them to return, on the eastern margin of a bend in the cove, on the eastern side of a creek, to shoot them, though invisible, by shooting across the head of the wake, which they made upon the water in alighting, and from which the few remaining rays of twilight that came from the western sky were reflected.

"But I am far from being singular in this. That page is more extensively read than is generally supposed. Many plain, unassuming men—farmers, shipmasters, and others within the circle of my acquaintance—know more, practically, of the weather than the most learned closet-theorist, or the most indefatigable recorder of its changes. Every one, by studying the page of nature above him, as he would the page of any other science, and testing, by observation, the numerous theories invented to account for the varied phenomena, may learn much, very much, that will be useful and interesting to him, and which he can never learn from books, or instruments, or theories alone."

"Well," said my friend, "I am too far advanced in life, as are many others, to commence such observations, and you must publish."

I demurred, and he insisted.

"It is difficult to spare the time; and I can not neglect my profession," I urged.

"Where there is a will there is a way," he replied.

"It is difficult to make one's self understood without many illustrations."

"Very well, they are easily obtained."

"But they cost money, and it is said 'science will not pay its way' like fiction and humbug."

"That," said he, "is a libel—such science will. Every one is interested in the weather—all talk about it—and thousands would carefully observe it, if they could be correctly guided in their observations."

"I may get into unpleasant controversy."

"Suppose you do; you can yield your position if wrong, and maintain it if right, and *magna est veritas*."

"But I may be mistaken in some of the views to which it will be necessary to advert, if I attempt to systematize the subject."

"Be it so—your mistakes may lead others to the discovery of the truth. Besides, the weather is *common property*, and every one has a right to theorize about it, or to talk about it, as they please—even to call a stormy day a pleasant one, or make any other mistaken remark concerning it; and every other person is entitled to a like latitude of reply. And further," said he, with some emphasis, "no important observation, in relation to a subject of such interest, should be lost; and, if you have observed one new fact, or drawn one new and just inference from those which have been observed by others; and especially if, from observation and reading, you can deduce from the phenomena an intelligible, *observable, general system*, it is not only your right, but duty, to make it known. Such a knowledge of the true system is greatly desired by every considerate man."

To my friend's last argument I was compelled to yield. I could make no reply consistent with the great principles of fraternity, which I shall ever recognize. The promise was given. My friend went on his way, and I went to the daguerreotypist to procure a copy of the then appearance of the sky, as the first step toward its fulfillment. The fulfillment of that promise, reader, you will find in the following work. It was commenced as an article for a magazine, but it has grown on my hands to a volume. Justice could not well be done to the subject in less space. It has been written during occasional and distant intervals of relaxation from professional avocations, or during convalescence from sickness, and it is, for these reasons, somewhat imperfect in style and arrangement. But I have no time to rewrite. There is much in it which will be old to those who read journals of science, but new to those who do not. There is more which will be new to all classes of readers, and may, perhaps, be deemed heretical and revolutionary by conservative meteorologists; yet I feel assured that the work is a step in the right direction—that it contains a substantially accurate exposition of the Philosophy of the Weather, and valuable suggestions for the practical observer.

I have inserted my name in the title-page, contrary to my original intention, and at the suggestion of others; for I have no scientific reputation which will aid the publisher to sell a copy. Nor do I desire to acquire such reputation. It can never form any part of my "capital in life." Nor has it influenced me at all in preparing the work. I have aimed to fulfill a promise, too hastily given, perhaps—to put on record the observations I have made, and the inferences I have drawn from those of others—to induce and assist further observations, and, if possible, of a *general* and *connected character*—and to impress those who may read what I have written with the belief, that *they will derive a degree of pleasure from a daily familiarity with, and intelligent understanding of, the "countenance of the sky," not exceeded by that which any other science can afford them.*

I have examined, with entire freedom and fearlessness (but I trust in a manner which will not be deemed censurable or in bad taste) the theories and supposed erroneous views of others, for, in my judgment, the advancement of the science requires it. Says Sir George Harvey, in his able article on Meteorology, written for the Encyclopædia Metropolitana:

> "It is humiliating to those who have been most occupied in
> cultivating the science of meteorology, to see an agriculturist or

a waterman, who has neither instruments nor theory, foretell the future changes of the weather many days before they happen, with a precision which the philosopher, aided by all the resources of science, would be unable to attain."

The admissions contained in this paragraph, in relation to the comparative uselessness of instruments and theories, and the value of practical observation, are both in a good measure true. And the time has come, or should speedily come, when *"pride of opinion,"* and *"esprit du corps,"* among theorists and philosophers, should neither be indulged in, nor respected; and when their theories should be freely discussed, and rigidly tested by the observations of practical men. Such measure, therefore, as I have meted, I invite in return. Let whatever I have advanced, that is new, or adopted that is old, be *as* rigidly tested, and *as* freely discussed. Let the errors, if there be any—and doubtless there are—be detected and exposed. Let the TRUTH be sought by all; and meteorology, as a PRACTICAL SCIENCE, advance to that full measure of perfection and usefulness, of which it is unquestionably susceptible.

CONTENTS

CHAPTER I.

Heat and moisture are indispensable to the fertillity of the earth—
Arrangements exist for their diffusion and distribution, and all
the phenomena of the weather result from their operation—Heat
furnished or produced mainly by the direct action of the sun's
rays—Manner in which it is diffused over the earth—Other causes
operate besides the sun's rays—The earth intensely heated in its
interior—Heat derived from the great Oceanic currents, and the
aerial currents which flow from the tropics to the poles, and from
magnetism and electricity—Water distributed by an atmospheric
machinery as extensive as the globe— Evidences of this—Its distribution over the continents of North America—Explanation of
it—Source from whence our supply of water is derived, and from
which our rivers return

CHAPTER II.

Our rivers return in the form of clouds, and in storms and showers—
Definition and character of storms—Differences in the character
of the clouds which constitute them— Nomenclature of Howard—Its imperfections—New order of description—Low fog—
High fog—Storm fog—Storm scud —N. W. scud—Cumulus—
Stratus—Cirrus—Compounds of the two latter—recapitulation in
tabular form

CHAPTER III.

Our rivers do not return from the North Atlantic—All storms and
showers move from the westward to the eastward— Seeming clouds seen moving from the eastward to the westward are

CHAPTER X.

CHAPTER XI.

THE PHILOSOPHY OF THE WEATHER.

CHAPTER I.

Heat and moisture are indispensable to the fertillity of the earth—
Arrangements exist for their diffusion and distribution, and all
the phenomena of the weather result from their operation—Heat
furnished or produced mainly by the direct action of the sun's
rays—Manner in which it is diffused over the earth—Other causes
operate besides the sun's rays—The earth intensely heated in its
interior—Heat derived from the great Oceanic currents, and the
aerial currents which flow from the tropics to the poles, and from
magnetism and electricity—Water distributed by an atmospheric
machinery as extensive as the globe— Evidences of this—Its distribution over the continents of North America—Explanation of
it—Source from whence our supply of water is derived, and from
which our rivers return

Heat and moisture are indispensable to the fertility of the earth. Without suitable arrangements for their diffusion and distribution, and within the limits of
certain minima and maxima, it would not have been habitable, or the design of
its Creator perfected. These arrangements therefore exist, and "while the earth re-
maineth seed time and harvest shall not cease." Few and simple in their character,
though necessarily somewhat complicated and irregular in their operation, the
ultimate result is always attained. A beautiful system of compensations supplies
the losses of every apparent irregularity in one section or crop, by the abundance
of others.

From the operation of these few, simple, connected, and intelligible arrangements for the diffusion of heat and the distribution of moisture over the earth,
result all the phenomena which constitute the weather; and by studying them, and
their operation, we may acquire an accurate knowledge of its *Philosophy.*

The necessary heat is furnished, or produced, mainly by the direct action of
the sun's rays; and the most obvious feature in the arrangements for its diffusion is
that by which the sun is made to shine successively and alternately upon different
portions of the earth. Nothing animate or organic could endure his burning rays,
if they shone continuously or vertically upon one point, or could exist without
their occasional presence. Hence the provision for a diurnal rotation, to prevent
the exposure of any portion of the globe to the action of those rays for twenty-four

consecutive hours, except for a limited period, and at a considerable angle, in the polar regions. But the earth is spheroidal, and a diurnal revolution would still leave that portion which lies under the equator too much, and the other too little, exposed to the action of the sun. This is obviated by an annual revolution of the earth around the sun, and an obliquity of its axis, by reason of which the northern and southern portions are alternately and, as far as the tropics vertically, exposed to the sun; and it is made to travel (so to speak) from tropic to tropic, producing summer and winter, and other important phenomena.

This obliquity and consequent change of exposure are in degree precisely what the wants of the earth would seem to require. If it was greater, the sun would travel further north and south, but the alternate winters would be longer and more severe. If it was less, the end would not be as perfectly attained.

The direct action of the sun's rays upon the earth, particularly those portions which lie north and south of the tropics, is not the only source from which the supply of heat is derived. Although there is a general increase of heat in spring and summer when the sun travels north, and of cold when he travels south in winter, yet there are frequent irregularities attending both. Very sudden and great changes occur in each of them. Frost sometimes, cool weather often, occurs in mid-summer, and considerable heat and tornadoes in midwinter. And ordinarily the maxima and minima of each month and, indeed, of each week are widely apart. Even in the polar regions, in midwinter, *where the sun does not shine at all*, the same moderating changes with which we are conversant occur in degree. An extract or two from the register found in Dr. Kane's narrative of the "Grinnell Expedition" will illustrate this.

JANUARY 1851, (LATITUDE ABOUT 74°, LONGITUDE ABOUT 70°).

Date.	Wind.	Force.	Ther.	Bar.	Sky and Weather.
Jan. 3	......	calm	-26.1	29.62	blue sky, m.
"4	W.	gent breeze	-21.3	29.53	blue sky, detached clouds, m.
"5	W. by N.	gent breeze	-3.9	29.59	blue sky, m., clouded over.
"6	W. by S.	light breeze	-0.8	29.67	clouded over, m., snow.
"7	W.	gent breeze	-14.4	29.96	blue sky, detached clouds, m.
"8	W.S.W.	light air	-21.2	30.14	blue sky, m.
"29	W.N.W.	light air	-18.9	30.19	blue sky.
"30	NW. by W.	light air	-13.5	30.17	clouded over, m.
"31	NW. by W.	gent breeze	-4.4	29.35	clouded over, snow.
Feb. 1	W.	light breeze	-11.7	29.27	cloudy, blue sky, m.
"2	W.	light air	-25.1	29.62	blue sky, detached clouds, m.

These extracts are instructive. It will be seen that on the 3d of January, when the sun had been absent some weeks, it was calm, the thermometer stood at 26° below zero (the - or minus mark before the figures indicates that), and the barometer at 29.62, with blue sky, somewhat misty or hazy—(the letter "m." standing for misty or hazy)—a state of the air which existed most of the time when it did not snow or rain, and therefore is of no importance in this connection. The next day the thermometer began to rise, and the barometer to fall. On the 5th it clouded over, and the thermometer rose rapidly, and on the 6th it had risen more than 25°, and snow fell. On the 7th it cleared off, the thermometer fell rapidly, and the barometer rose. On the 8th the thermometer had fallen to 21° below zero, and the barometer had risen to 30.14. Another instance, in all respects similar, occurred the latter part of the month. We shall see hereafter that these changes are precisely like those which occur with us, and every where. That, as in the polar regions, and whether the sun be present or absent, or obscured by clouds, and by night as well as by day, the changes from warm to cold and from cold to warm are sudden and great, and that the latter are connected with the fall of rain and snow—that every where in winter it *moderates to storm.*

Many other instructive instances, especially in relation to the great difference in the seasons in our own country, and upon the same parallels elsewhere, might be cited if it were necessary. But they will more appropriately appear in the sequel.

The cause of those irregularities, especially in the same seasons of different years, and when very great, is often sought and supposed to be found in the presence or absence of spots on the sun, ice floes and bergs in the Atlantic, etc., etc. But neither the spots, nor ice, nor other local causes produce them. The cause will be found in the character of the arrangements we are considering, and the irregular action of the power which controls them.

Nor is the temperature of the northern hemisphere, north of the tropics, equal in the same latitudes. Very great diversities exist in the "annual mean" as well as the "mean" of the different seasons. Accurate observations at many points have enabled men of science to demonstrate this by drawing isothermal lines (*i. e.*, lines of equal average annual heat) from point to point around the earth, which show at a glance these differences. The annexed cut is a polar projection of the isothermal lines of the northern hemisphere, as far down as the tropic, copied from Kaemtz's Meteorology. The dotted lines show the parallels of latitude, the dark lines the isothermal lines, or lines of equal annual average temperature. The reader is desired to observe how rarely they correspond with the parallels of latitude, and how they fall below in a few instances, and in others with great uniformity rise almost to the pole.

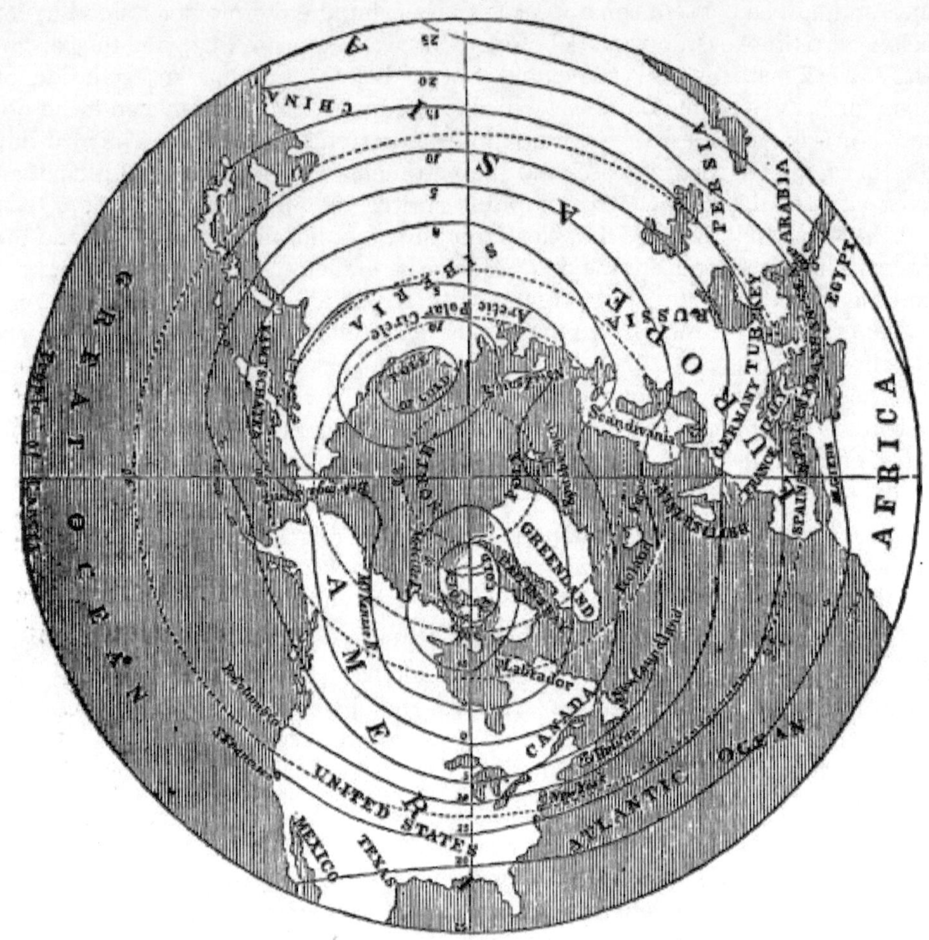

FIG. 1.

In the above cut the isothermal lines are Centigrade. The zero of the Centigrade thermometer is the freezing point of water, or 32° of Fahrenheit. The boiling point of water is 100° Centigrade, or 212° Fahrenheit. A degree of Centigrade is equal to one degree and four-fifths, Fahrenheit. The 0° line of the cut, therefore, is 32° of Fahrenheit—the line of 5° above is 41° Fahrenheit—the line of 5° below is 23° Fahrenheit, and so on. The reader, who is not familiar with the difference in the scale of the thermometer, is desired to remember this; for we shall make occasional extracts in which the temperature is given in the Centigrade scale.

Take, for example, the isothermal line of 0 or zero—that is, the line where the mean or *average* height of the thermometer *for the year* is at zero. At Behring's Straits this line is a little below the Arctic circle, or the parallel of 66.30 north latitude. Passing east over North America, it descends into Canada, almost to Lake

Superior, and to about the 50th parallel: that is to say, it is on an average during the year as cold on our continent at the 50th parallel as it is near Behring's Straits at the 65th parallel. Passing east, the line of zero rises again over the Atlantic Ocean until, in the meridian of Spitzbergen, it reaches, within the Arctic circle, up almost to the 75th parallel. So, too, the isothermal of 5° below zero, which is below the 60th parallel in Siberia, rises in the North Sea, above Behring's Straits, to the parallel of 75°, descending on the continent in North America to the 55th parallel, and rising again almost to the pole at Spitzbergen, to descend again in Siberia, while the isothermals of 10° and 15° below zero, which in North America are but just above the latitude of 60° and 75° respectively, ascend abruptly *surrounding the magnetic pole,* and *falling short of the geographical one.* Let this projection of the lines of equal temperature, and particularly the situation of the magnetic poles, be studied well, for we shall recur to it hereafter in illustration of many important portions of our subject.

It is apparent from these facts, and were it necessary might be rendered still more so by referring to others, that other causes operate in the distribution of heat over the earth besides the direct action of the sun's rays upon it. Doubtless very considerable allowance is to be made for the difference of seasons, and difference during the same season upon the land and upon the ocean; in mountainous countries and level ones. But making every allowance for them, the fact that other causes have a *controlling* influence in producing the deviations still remains most obvious. Neither the difference of temperature between the land and the ocean, or land surfaces of unequal elevations, will account for the elevation of the isothermal lines on different portions of the ocean, or their extension around the magnetic poles.

Returning to a consideration of the arrangements for the diffusion of heat, we observe: First, that the earth itself is intensely heated in its interior. This is inferred, and justly, from the fact that the thermometer is found to rise about one degree for every fifty-five feet of descent—whether in boring artesian wells, exploring caves, or sinking shafts in mines. It is demonstrated, also, by the existence of hot springs and the action of volcanoes. Heat is supposed to be conducted from the center toward the surface every where, but with difficulty and slowly. It is also supposed to be conducted from the tropical regions toward the poles. Such is the opinion of Humboldt. (Cosmos, vol. i. p. 167.)

Probably it reaches the surface and exerts an influence, also, upon the weather through the ocean, and by heating it in its greatest depths. Little attention has been paid, so far as I am informed, to the question how far the ocean is thus heated in *tropical latitudes.* Doubtless a portion of the warmth of the ocean there is derived from that source, and it has its influence in changing the temperature of the deep-seated cold polar currents of, the great oceans. Perhaps it may yet be found that the icebergs are detached by it in the polar seas—the observations of Dr. Kane point to such a result. (Grinnell Expedition, p. 113, and also chap. 48.)

Little need be said of the inconsiderable quantities of heat supposed to be derived by radiation from the stars, the planets, and from space. If any such are derived they are too inconsiderable to be of importance in this inquiry.

Heat is also carried, and in quantities which exert very considerable influence upon the weather, from the tropics to the poles by the great oceanic currents which flow unceasingly from one to the other.

The most important of these with which we are acquainted is the Gulf Stream of the Atlantic. Gathering in the South Atlantic, and passing north through the Caribbean Sea and the Gulf of Mexico, it issues out through the Bahama Channel, and flows north along the eastern coast of the United States, but some distance from it, to Newfoundland, and from thence continuing to the north-east and spreading out over the surface of the ocean—a portion of it mingling with the waters of the North Atlantic in passing—it flows up on the western coast of Europe, around the Faroe Islands, and Spitzbergen, to the polar sea; passing around Greenland, and perhaps through its Fiords, it descends again through the sounds and channels of the Arctic regions into Baffin's Bay, and through Davis's Straits, burdened with the icebergs and floes of the polar waters, to return again to the South Atlantic. For reasons which will appear in the sequel, it has comparatively little influence upon the weather of the United States. Western Europe, however, Greenland, the islands which lie in its course, and the polar seas, are most materially influenced. Although not the only cause, it has very much to do with the remarkable elevation of the isothermal lines over the Northern Atlantic, and upon Western Europe, as seen upon the map.

A like oceanic current exists in the Pacific Ocean, the influence of which may also be traced upon the map by the elevation of the isothermal lines at the northern extremity of that ocean, and upon the north-west coast of North America. A vast amount of heat is transported from the tropical to the temperate and frozen regions of the earth by these great oceanic currents.

Another supply is derived from aerial currents which flow from the tropics toward the poles. These currents exist every where over the entire surface of the earth, but in more concentrated volumes along the great "lines of no variation," and greater magnetic intensity, on the western side of the great oceans, over the eastern portions of the two continents of North America and Asia. Not, as meteorological writers suppose, in the upper portions of the atmosphere, having risen in the trade-wind region and run off at the top toward the poles by force of gravity, but near, and sometimes in contact with the earth. The influence of these aerial currents upon the temperature of the atmosphere, and in producing the phenomena we are to consider, is exceedingly important. We shall have occasion to examine them with great care and minuteness under another head, for upon them, more than any other portion of the arrangements, depend not only the diffusion of heat, but also the distribution of moisture.

Still another supply of heat, during the sudden changes, at least, is produced by the action of terrestrial magnetism and electricity. Very great progress has been made within a short period, in the investigation of the nature of these agents. The identity, or at least intimate association or connection of heat, light, electricity, and magnetism, always suspected, has been in various ways, and by a variety of experiments demonstrated. The influence of magnetism if distinct from gravitation, is second only to that; and its agency in producing the phenomena we are considering is primary and controlling. We will only, in this connection, ask the reader to note the situation of the north magnetic poles (for there are two of them); the manner in which the isothermal lines *surround* them; the fact that they are *poles of cold*, *i. e.*, that it is colder there than even to the north of them. We shall recur to this part of the subject again.

Such, briefly considered, are the principal arrangements by which heat is diffused over the earth.

Equally marked by infinite wisdom, and equally interesting and important, are the arrangements by which moisture is distributed. Doubtless the general belief is that this is a simple process; that water evaporates and rises till it meets a colder stratum of atmosphere, and then condenses and falls again; or that, according to the Huttonian theory, currents of air of different temperatures mingle and equalize their heat, and the aggregate mass when equalized in temperature is cooler, and therefore is unable to hold as much moisture in solution as the most heated portion had, and the excess falls in rain. But the process is by no means so simple, nor is heat the sole or most powerful agent concerned in it. Currents of air do not mingle, but stratify. Evaporation from the surface of any given portion of the earth outside of the tropics does not alone supply that portion with rain. *Vast and wonderful, coextensive with the globe itself, and perfectly connected, is the machinery by which that supply is furnished even to the most inconsiderable portion of its surface.*

Take your map of North America and note, in this respect, its peculiarities. It extends from the Isthmus of Darien to the Arctic regions, and from the 65th to the 160th meridian of west longitude from Greenwich, and has upon its surface a type of every climate in the world. For the purpose of simplifying and illustrating the matter in hand, let us divide it into five sections. Let the first section embrace Central America and Southern Mexico, south of 28°; the second, Northern Mexico and Southern New Mexico, California, etc., between the parallels of 28° and 32°; the third, Northern California, Utah, Southern Oregon, and Western New Mexico, north of the parallel of 32°; the fourth, the entire continent north of 42°; and the fifth, the eastern United States, east of the meridian of 100°. These divisions are not intended to be entirely accurate in their separation, but substantially so for the purpose of illustrating the differences which exist in each.

The accompanying diagram shows approximately, by dotted lines, the divisions.

North Pole
GREENLAND
M.Kenzie R.
BRITISH
4TH
AMERICA
UNITED
3D
Missouri R.
Mississippi
St. Lawrence
STATES
5TH
2ND
MEXICO
GULF OF
MEXICO
1ST

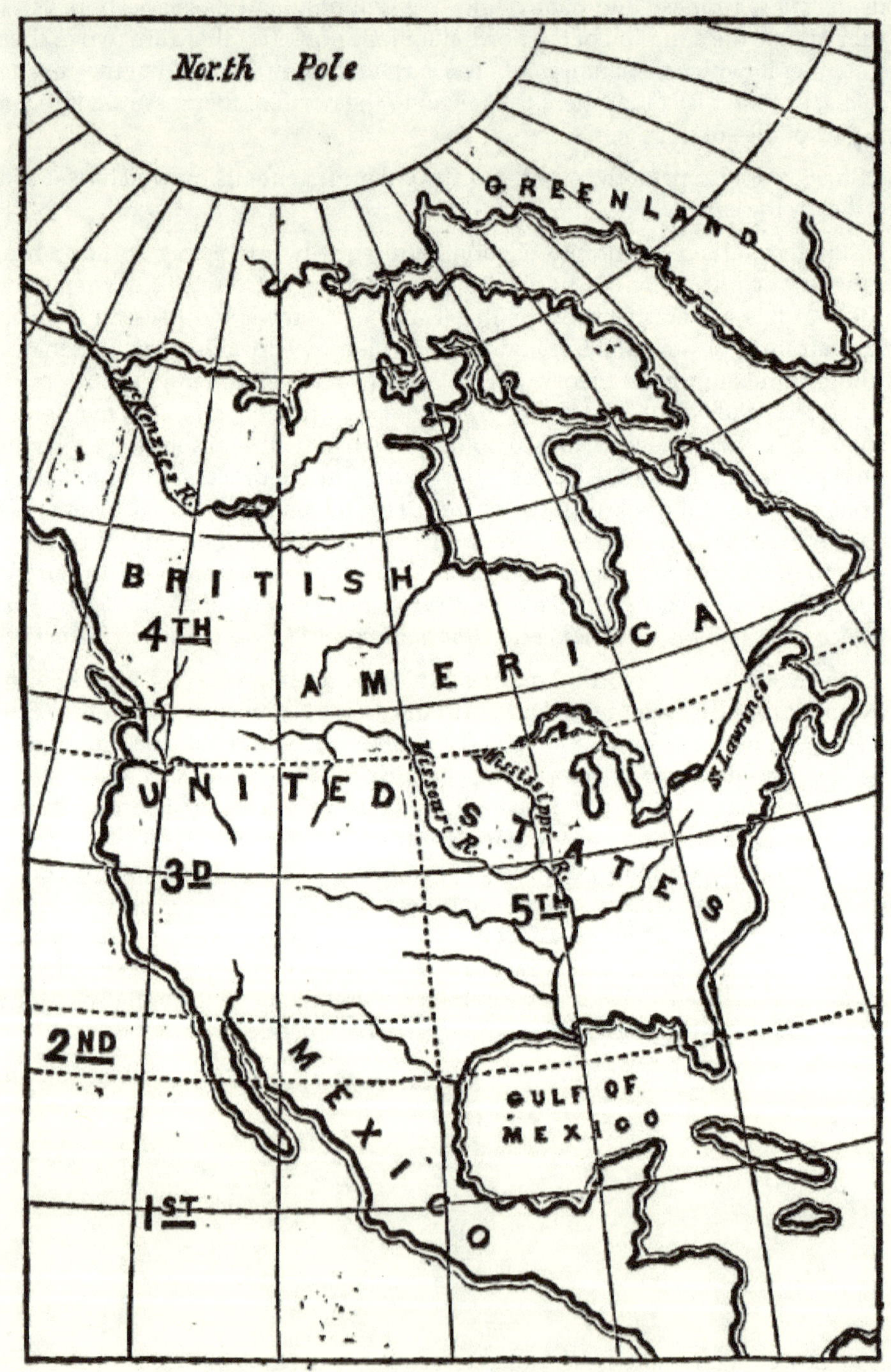

FIG. 2.

Now let us see in what a diverse manner, and to what a different extent, they are severally supplied with moisture.

Central America and Southern Mexico lie within the tropics—their rains are tropical rains. The season is divided into wet and dry, as are the seasons of all tropical countries which are not rainless. During the rainy season it rains a portion of nearly every day, and during the dry season the sky is clear, the air is pure, and rain seldom falls.

All around the earth within the tropics, over the land and over the sea, there is a belt of almost daily rains, varying in width, north and south, in different sections, but averaging about five hundred miles. This belt of daily rains is formed at and by the meeting of N. E. and S. E. trades, and travels north and south with them, as they do with the sun, *encircling the globe.* By this narrow belt a portion of the earth's surface, an average of some 35° of latitude, is supplied with moisture. Wherever it is situated at any given period, the tropical rainy season exists; and when it is absent in its northern or southern transit, the dry season prevails. Southern Mexico is within the range of this moving belt, and in its course to the northward with the sun, in our summer from May to October, it arrives over, and covers that country with a rainy season. When the sun returns to the south, taking with it the trades and this belt of tropical rains, that portion of Mexico is without rain, and dry, and so continues until the rainy belt returns in the following year. While the belt is over Southern Mexico it is nearly all *precipitation,* and there is little *evaporation;* while that belt is *absent* it is all *evaporation,* with little or no *rain.* Surely this is not consistent with the prevailing belief of simple evaporation, ascent to a colder stratum, commingling, and condensation, and rain. Southern Mexico at least is not supplied by mere evaporation from its surface, and must therefore form an exception to that belief, and to the Huttonian theory.

But we shall recur again to the peculiarity of distribution within the tropics.

Turn now for a brief space to Northern Mexico, Southern New Mexico, and Southern California. In Northern Mexico, Southern New Mexico, Utah, and California, between the parallels of 28° and 32°, and particularly west of the mountain ranges, we find an almost rainless region, sterile and worthless, resembling that which is found upon nearly the same parallels of north latitude in Northern Africa, Egypt, Arabia, Beloochistan, Afghanistan, and North-western India; and in corresponding latitudes south of the Equator, in Peru, a portion of Southern Africa, and the northern and middle portions of New Holland. Why Northern Mexico and the other countries named are thus sterile and comparatively rainless, we shall see hereafter, when we examine critically the machinery of distribution as it operates within the tropics. It is the fact that it is thus sterile and rainless to which we desire to call attention in this place.

Mr. Bartlett thus describes it:

> "On leaving the head waters of the Concho, nature assumes
> a new aspect. Here shrubs and trees disappear, except the thorny
> chaparral of the deserts; the water-courses all cease, nor does any
> stream intervene until the Rio Grande is reached, three hundred

and fifty miles distant, except the muddy Pecos, which, rising in the Rocky Mountains, near Santa Fé, crosses the great desert plain west of the Llano Estacado, or Staked Plain.

"From the Rio Grande to the waters of the Pacific, pursuing a westerly course along the 32d parallel, near El Paso Del Norte, there is no stream of a higher grade than a small creek. I know of none but the San Pedro and the Santa Cruz—the latter but a rivulet, losing itself in the sands near the Gila—the other but a diminutive stream, scarcely reaching that river. At the head-waters of the Concho, therefore, begins that great desert region, which, with no interruption save a limited valley or bottom-land along the Rio Grande, and lesser ones near the small courses mentioned, extends over a district embracing sixteen degrees of longitude, or about a thousand miles, and is wholly unfit for agriculture. It is a desolate, barren waste, which can never be rendered useful for man or beast, save for a public highway."—*Bartlett's Personal Narrative*, vol. i. p. 138.

Turning now to Central and Upper California, and Utah, and Southern Oregon, we find still another peculiarity. Like Southern Mexico, they have a rainy and dry season, but at a different period, and for a different reason. The dry season of California, etc., is the summer of the northern hemisphere, and her rainy season the winter. *California* is, therefore, *dry* when Southern *Mexico* is *wet*, and *vice versâ*. The belt of rains which supplies California with moisture during her rainy seasons is the belt of *extra-tropical* rains, which extends from the northern limit of the north-east trades to the poles, encircling the earth. The southern edge of this extra-tropical belt is *carried up* on the western coast of America, and in that portion of the continent in *summer*, when the sun and trades, and the inter-tropical rainy belt travel to the north, and uncover California, etc., leaving them without rain for a period of about six months.

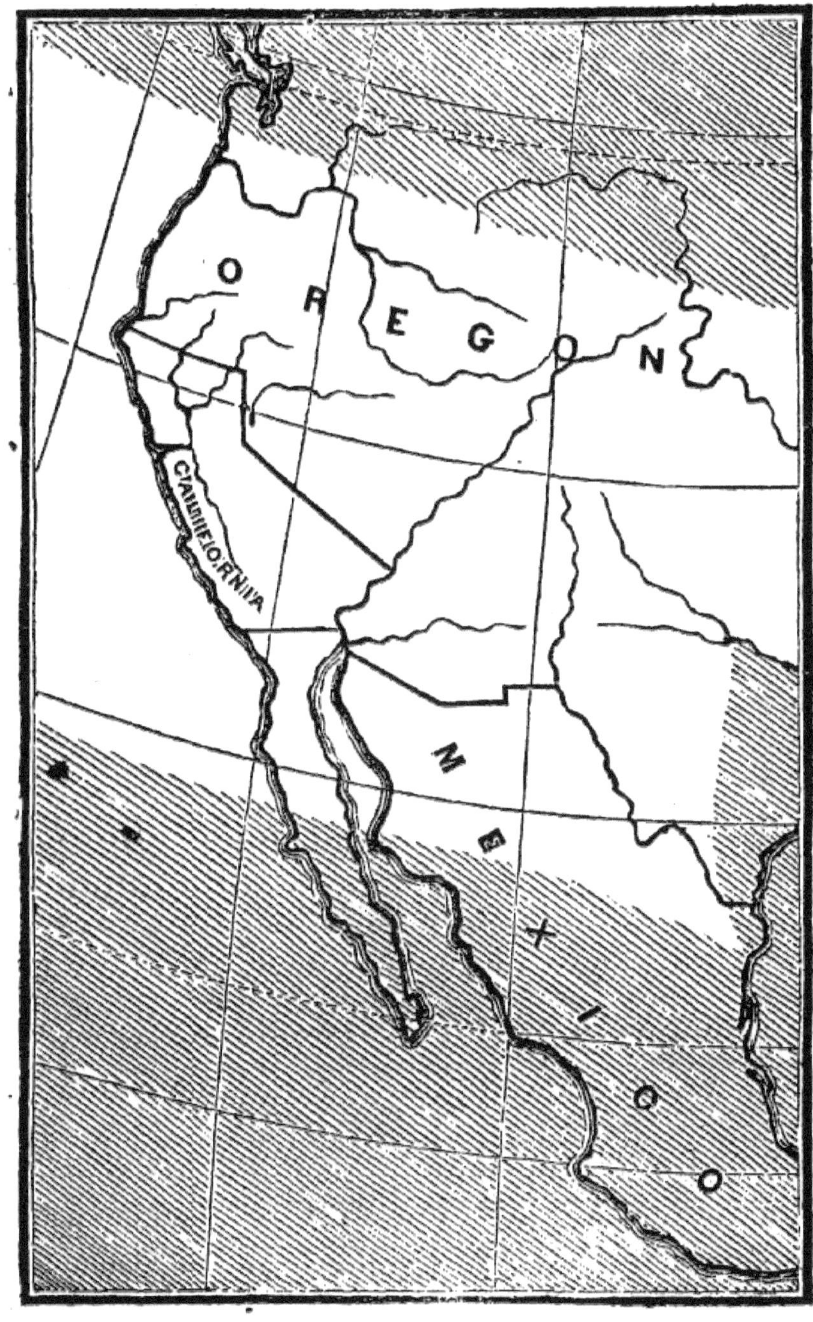

FIG. 3.

IN SUMMER.

As the sun, with the trades, travels south, the southern edge of the belt of extra-tropical rain follows, and covers California, etc., again extending gradually from the north to the south, and thus their wet season returns. The annexed diagrams by the shading will show the situation of the rainy belts which cover Mexico, Utah, New Mexico, and California in summer and winter, and that the belts of rains are entirely distinct and different in character.

FIG. 4.
IN WINTER.

Here again in this section of the continent, as in Mexico, evaporation is going on for six months of the year, and were it not for the return of the belt of rains from the north, in the fall, would go on for the entire year without precipitation; and for the other six months precipitation is vastly in excess. Nor can this be reconciled with, or explained by, the Huttonian or any other received theory of rain. Here again it is obvious that evaporation alone, however great or long continued, will not furnish the evaporating section with rain.

The northern portion of the continent lies beneath the zone of extra-tropical rains, and north of the northern limit of the N. E. trades—is never uncovered from it, and has no distinct rainy or dry season, although more rain falls at certain periods, and in certain localities, than at others. The climate of that part of Oregon which lies upon the Pacific, and the character of its rains, resemble those of North-western Europe, and will be further explained hereafter.

Coming to the portion of the continent which we occupy, the 5th section, we find it different still—a most favored region. Portions of it—Eastern Texas, for instance—are upon the same parallels of latitude as the rainless regions of Northern Mexico, etc. Eastern Texas, however, is not rainless. Other portions are upon the same parallels as California, etc., yet have no distinct rainy and dry season. We repeat, this section is a most favored region—without a parallel upon any portion of the earth's surface, except, in degree, in China and some other portions of Eastern Asia.

It is not only without a distinct rainy and dry season, but it is watered by an average, annually, of more than forty inches of rain, while Europe, although bounded on three sides by seas and oceans, and apparently much more favorably situated, receives annually an average of only about twenty-five—if we except Norway, and one or two other places, where the fall is excessive. The distribution of this supply of moisture over the United States is, in other respects, wonderful. Iowa, in the interior of the continent, far away from the great oceans, on the east or west, or the Gulf of Mexico on the south, receives fifty inches; some ten or fifteen inches more than fall upon the slope east of the Alleghanies, and contiguous to the great Atlantic (from which all our storms are, erroneously, supposed to be derived), and the average over the entire great interior valley is about forty-five inches, falling at all seasons of the year.

Observe, then, by way of recapitulation: Southern Mexico has a rainy season furnished by the belt of *inter*-tropical rains, which *travels up over it from the south* in summer. California has a rainy season, which is furnished by the *extra*-tropical belt of rains, which travels *down from the north*, and covers it in winter. Northern Mexico and the adjoining regions west of the 100th meridian are between the limits of the two, and neither travels far enough to reach them, except for brief and uncertain periods; they are comparatively rainless; while the eastern portion of the continent, *in all latitudes*, unlike the others, is without a distinctly marked dry season, or a rainless region, and with the exception of occasional droughts, is abundantly supplied with rain at all seasons of the year.

And now, what is the explanation of all this? What produces the extra-tropical belt of regular rains surrounding the earth, north of the parallel of 30° north, in some places, and 35° in others, extending to the pole, with its southern edge traveling up ten or more degrees in summer, leaving large portions of the earth subject to a dry season; and back again in the winter to give them a rainy one? What produces the narrow belt of inter-tropical rains, encircling the earth; traveling up and down every year over an average of 35° of latitude, supplying every portion of it alternately with rain? And what connects the two together over the eastern portion of North America, so as to leave no distinctly marked wet and dry season, and no rainless and sterile portion there? Are all these the result of simple evaporation, ascent to a colder region, condensation, and descent again? Demonstrably not. Of the forty inches which fall annually upon the middle and eastern portions of the United States, an average probably of one-half or twenty inches, runs off by the rivers to the ocean, or is carried away eastward by the westerly and north-westerly evaporating winds. The same is true, in degree, of the rain which falls upon the other portions. Evaporation, therefore, could not keep up the supply. From whence, then, does it come? this twenty inches, thus lost by the rivers and winds, and with such wonderful regularity every year.

"All the rivers run into the sea, yet the sea is not full. *Note the place whence the rivers come, hither they return again.*"

But how is it that they thus return with such wonderful regularity, in a narrow traveling belt of daily rains within the tropics, and a movable belt of irregular rains without the tropics, extending to the poles, leaving a space on each side of the equator encircling the earth in like manner (except at two points, *viz.*, Eastern Asia and Eastern North America), from which they do not go, and to which they do not return, and which is almost entirely unfurnished with rain? And all this without any relation, whatever, to the contiguity of the oceans? Obviously this is not the work of mere evaporation, or of the accidental or irregular commingling of winds with different dew points, or quantities of moisture in solution, or accidental, irregular changes of barometric pressure. *It is one vast, wonderful, connected, and regular system—co-extensive with the globe—necessary to the return of moisture from the oceans upon the most inconsiderable portion of it, and to the condensation of the local moisture of evaporation; and by it the waters are returned from the oceans as regularly and bountifully upon the far interior of the great continents in the same latitudes, as upon the "isles which rest in their bosoms."*

CHAPTER II.

Our rivers return in the form of clouds, and in storms and showers—
Definition and character of storms—Differences in the character
of the clouds which constitute them— Nomenclature of Howard—Its imperfections—New order of description—Low fog—
High fog—Storm fog—Storm scud —N. W. scud—Cumulus—
Stratus—Cirrus—Compounds of the two latter—recapitulation in
tabular form

Before proceeding to an examination of this connected atmospheric machinery, and an investigation of the particular ocean from which our rivers return, it
may be well to look at the form in which they appear to return, that we may have
a clear understanding of terms.

They seem to return in the form of clouds, and in storms and showers, although, in truth, they return in regular, uniform, ordinarily invisible currents, and
the storms and showers are but condensations in, and discharges from portions of
those currents, aided by the local moisture of evaporation.

The term *storms,* seems to be used by European meteorologists to denote what
we term thunder showers or gusts, and tornados; while what we call storms are
denominated by them regular rains. As the terms are extensively in use in this
country, we must adhere to the meaning attached to them *here* rather than *there.*

Storms with us, then, are regular rains of from six to forty-eight or more hours'
continuance: generally without lightning, or thunder, or gusts, and usually with
wind of more or less force, from some easterly point. They are called north-east
storms, or south-east storms, according to the point from which the surface winds
blow. Practically we shall find that this distinction is of some importance, for the
north-east storms are the longest, lasting generally twenty-four hours, or more,
while the south-east ones seldom, if ever, continue as long.

These storms extend over a considerable surface, rarely less than one hundred miles in one direction or another, and sometimes fifteen hundred, or more.
Distinct showers cover but a small surface, sometimes not more than forty to one
hundred rods, as in the tornado, and rarely more than ten miles. Belts of showers,
each new one forming a little more to the south, often, in summer, pass across the
country, following each other in succession; and these belts may be of considerable
width, say thirty to one hundred and fifty miles.

The clouds which constitute the storms and showers differ in appearance and
character, as well in the active as in the forming state. Clouds are of distinct char-

acters, alike, substantially, every where under like circumstances; and a distinct nomenclature has been applied to them by Dr. Howard, of London. He notes three kinds of primary clouds: *viz.*, cirrus, stratus, and cumulus; and inasmuch as the boundary line between them is not very distinct, certain compounds of the three, *viz.*: cirro-stratus, cirro-cumulus, and cumulo-stratus. This nomenclature is every where received, and portions of it are of great practical importance.

The three principal descriptions of cloud, *viz.*: the cirrus, the stratus, and the cumulus, we have very much as they have in Europe, and doubtless as they exist every where outside of the tropics. The nimbus, another cloud described by him, is not distinct from the cumulus or stratus. An isolated, limited thunder-shower in a clear sky, presents the appearance of a nimbus, as shown in the cuts, but the basis of it is a cumulus, and it differs from an ordinary fair-weather cumulus merely in the dark and fringe-like appearance of the rain as it is falling from its lower surface, and sometimes in the existence of a stratus above and in connection with it. A similar form is often assumed by the peculiar clouds of the N. W. winds in March or November, when they assume the form of *squalls*, and drop flurries of snow. The nimbus, therefore, is not a distinct cloud, but an appearance which the cumulus, stratus, or cirro-stratus has in a stormy or showery state, and does not deserve a distinct name. It is but a cumulus, or a stratus, or cirro-stratus dissolving in snow or rain. It is important that this term should be abandoned. It tends to confuse and prevent a clear understanding of the difference in the character of the clouds, and in relation to which precision is both difficult and desirable.

The figures on pages 27 and 29, show the different kinds of clouds as designated by Howard. They are copied from the engravings in the sixth edition of Maury's "Sailing Directions."

FIG. 5.

LARGER IMAGE

FIG. 6.
LARGER IMAGE

Figure 5.

The cirrus is indicated by	1 bird.
The cirro-cumulus by	2 "
The cirro-stratus by	3 "
The cumulo-stratus by	4 "

Figure 6.

The cirrus by	1 "
The cumulus by	2 "
The stratus by	3 "
The nimbus by	4 "

How far these representations correspond with the actual appearance of the different compound forms in England, I can not say. But although they convey a *general* idea, *they are not sufficiently accurate for practical illustration or observation here*. Indeed Howard himself has omitted from his last edition his plate of the clouds, assigning as a reason, "that the real student will acquire his knowledge in a more solid manner by the observation of nature, without the aid of drawings, and that the *more superficial are liable to be led into error by them*." The collection of forms in the cuts *does not contain some very important ones*, and contains some which are not distinct forms; but they may aid us somewhat in this inquiry, and, therefore, I have copied them. It is well, also, for the reader to have the generally received description before him.

But for the purpose of *practical* illustration hereafter, and greater precision, I shall follow a somewhat different order in describing them, and introduce two forms of *scud* quite as important, practically, as any other.

First, then, commencing at the earth, we have what may be properly termed *fog*, or low fog. This forms, in still clear weather, in the valleys, and over the surface of the rivers and other bodies of water, during the night, and most frequently the latter part of it, and is at its acmé at sunrise, or soon after, limiting vision horizontally and perpendicularly, and dissolving away during the forenoon. It is rarely more than from two to four hundred feet in height at its upper surface, and often much less, and is composed of vesicular condensed vapor, sometimes sufficiently dense to fall in mist, and is doubtless in composition substantially what the clouds are in the other strata of the atmosphere, as observed by us, or passed through by aeronauts. I have never seen it carried up to any considerable height into the other strata by any of the supposed ascending currents, to form permanent clouds, and shall have occasion to allude to the fact in another connection. It disappears usually before mid-day, and has, when thus formed, no connection with any clouds which furnish rain.

To this Dr. Howard originally gave the name of stratus, and so it is represented upon the cut; but the latter term may be with greater propriety applied to the smooth uniform cloud in the superior strata from which the rain or snow is known to fall, and I shall retain and so apply it.

The next in order, ascending, is high fog. This is usually from one to two thousand feet in height at its lower surface. It forms, like low fog, during the night and in still weather; and is rarely, if ever, connected with clouds which furnish rain. It breaks away and disappears between ten and twelve in the forenoon, usually passing off to the eastward. This fog is most commonly seen in summer and autumn, particularly the latter, and unless distinguished from cloud will deceive the weather-watcher. It is readily distinguishable. Although often very dense, obscuring the light of the sun as perfectly as the clouds of a north-east storm, it differs from them. It forms in still clear weather, is present only in the morning, is perfectly uniform, and, before its dissolution commences, without breaks, or light and shade, or apparent motion, and unaccompanied by scud or surface wind. The storm clouds are never entirely uniform, or without spots of light and shade, by which their nature can be discerned, and rarely, when as dense as high fog, without scud running under them and surface winds.

There is another fog still, connected with rain storms, but it does not often precede them; occurring at all seasons, but most commonly in connection with the warm S. E. thaws and rains of winter and spring; and which usually comes on *after* the rain has commenced and continued for awhile, and the easterly wind has abated; occupying probably the entire space from the earth to the inferior surface of the rain clouds or stratus. Practically this does not require any further notice. It is an *incident* of the storm. When formed it remains while the storm clouds remain, and passes off with them. It is sometimes exceedingly dense in February and March, when it accompanies a thaw, and if there is a considerable depth of snow, it has the credit of aiding essentially in its dissolution.

Mingled with the smoke of London, it produced there the memorable *dark day* of the 24th of February, 1832, and at various other times has produced others of like character. (See Howard's Climate of London, vol. iii. pp. 36, 207, 303.) These fogs have been so dense there that every kind of locomotion was dangerous, even *with lanterns, at mid-day*.

The next in order, ascending, are the storm scud, which float in the north-east or easterly, south-east or southerly wind, before and during storms.

These, as the reader will hereafter see, are, *practically*, very important forms of cloud condensation—although they have found no place in any practical or scientific description given of the clouds, and are not upon the cuts. They are patches of foggy seeming clouds of all sizes, more or less connected together by thin portions of similar condensation, often passing to the westward, south-westward, north-westward, or northward with great rapidity. Their average height is about half a mile, but they often run much lower. They are usually of an "ashy gray" color. The annexed cut shows one phase of them, from among many taken by daguerreotype. The arrows pointing to the west show the scud distinguished from the smooth partially formed stratus above. This view was taken a few hours prior to the setting in of a heavy S. E. rain storm. It is a northerly view.

FIG. 7.

At about the same height, but in a *different state of the atmosphere*, float the peculiar fair-weather clouds of the N. W. wind. They usually form in a clear sky, and pass with considerable rapidity to the S. E. Sometimes they are quite large, approaching the cumulus in form, and white, with dark under surfaces, and at others, in the month of November particularly, are entirely dark, and assume the character of squalls and drop flurries of snow; and then resemble the nimbus of Howard. They assume at different times and in different seasons, different shapes like those of the scud, the cumulus, or the stratus.

FIG. 8.

They form and float in the peculiar N. W. current which is usually a fair-weather wind, and are never connected with storms. In mild weather they are usually white, and in cold weather sometimes very black, and at all times differ *in color* from the ashy gray scud of the storm. This variety is not represented upon the general cuts. The annexed diagram shows one phase of them, but they are readily observable at all seasons of the year, when the N. W. wind is prevailing; differing in appearance according to the season. Let these, as well as the storm scud, be carefully observed and studied by the reader, and let no opportunity to familiarize himself with their appearance be lost. A brief glance at each recurrence of easterly or north-westerly wind will suffice.

SUMMER CUMULI.

The *cumuli* appear in isolated clouds of every size, or in vast clouds composed of aggregated masses, as the peculiar cloud of the thunder shower. They form as low down as the scud or fair-weather cloud of the N. W. wind, which, for convenience, I will call N. W. *scud*; and often in violent showers, and particularly in hail storms, extend up as far as the density of the atmosphere will permit them to form. Professor Espy thinks he has measured their tops at an altitude of ten miles. Others have estimated their height, when most largely developed, at twelve miles; but it is very doubtful whether the atmosphere can contain the moisture necessary to form so dense a cloud at that elevation. It is their immense height, however, whether it be six, or eight, or ten miles, together with the sudden and violent electric action, condensing suddenly all the moisture contained in the atmosphere within the space occupied by the cloud, which produces such sudden and heavy falls of rain or hail. As the rain drops or hail, when formed at such an elevation, in falling through the partially condensed vapor of the cloud must necessarily enlarge by accretion from the particles with which they come in contact, and probably also by attraction, their size when they reach the earth, though frequently very considerable, is not a matter of astonishment. The cumulus is represented in the general plate with sufficient accuracy to show its peculiar character.

In summer, when the air is calm, the weather warm, and no storm is approaching, there is always, in the day time, a tendency to the formation of cumuli. This tendency exhibits itself about ten o'clock in the forenoon, and they gradually form and enlarge until about two in the afternoon; and after that, if they do not continue to enlarge and form showers, they melt away and disappear before nightfall. Sometimes in July and August the atmosphere will be studded with them at mid-day, floating about three-quarters of a mile from the earth (in a level country), gently and slowly away to the eastward. At times it may seem as if they must coalesce and form showers, yet they frequently do not, but gradually melt away, as before stated.

The cumulus is the principal cloud of the tropics, and is not often seen with us except in summer, or when our weather is tropical in character.

The engraving on the preceding page, shows a phase of these fair-weather summer cumuli.

The last in order occupying (with their compounds) the higher portions of the atmosphere, are the cirrus and stratus. The cirrus is often the skeleton of the other, and precedes it in formation.

These are the proper clouds of the storm, in our sense of the term. While, however, the cirrus remains a cirrus, it furnishes no rain. When it extends and expands, and its threads widen and coalesce into cirro-stratus and stratus, or it induces a layer of stratus below it, the rain forms.

The following is Dr. Howard's description of cirrus: "Parallel, flexuous or diverging fibers, extensible by increase in any or in all directions. Clouds in this modification appear to have the least density, the greatest elevation, and the greatest variety of extent and direction. They are the earliest appearance after serene weather. They are first indicated by a few threads penciled, as it were, on the sky. These increase in length, and new ones are in the mean time added to them. Often the first-formed threads serve as stems to support numerous branches, which in their turn, give rise to others."

The illustrations in the general cut are imperfect, and do not represent the delicate fibers of the cloud, for it is a difficult cloud to daguerreotype or engrave, but the representation is sufficiently accurate to give the reader a general idea of the different varieties, and enable him to discover them readily by observation. They are the most elevated forms, always of a light color, and often illuminated about sunset by the rays of the sun shining upon their inferior surface; the sun, however, often illuminates, in like manner, the dense forms of cirro-stratus, and the latter, from their greater density, are susceptible of a brighter and more vivid illumination.

The stratus is a smooth, uniform cloud—the true rain cloud of the storm; often forming without much cirrus above, or connected with it. It may be seen in its partially formed state in the bank in the west, at nightfall, or in the circle around the moon in the night. When it becomes sufficiently condensed, rain always falls from it, but in moderation. If there be large masses of scud running beneath it for its drops to fall through (especially as is sometimes the case, in two or more currents), the rain may be very heavy. But more of this hereafter.

FIG. 10.

The annexed cut shows the forming stratus, light and thin, passing to the east, as indicated by the short arrows just before a storm, while the scud beneath is running to the west.

It was copied from a daguerreotype view, facing northwardly.

Intermediate between the fibrous, tufted, cirrus, and the smooth uniform stratus, there is a variety of forms partaking more or less of the character of one or the other, and termed *cirro-stratus*. No single correct representation of cirro-stratus as a distinct cloud, can be given—but several varieties will be hereafter alluded to, under the head of prognostics. Several modifications are represented with tolerable accuracy upon the cuts.

The cirro-cumulus is a collection in patches of very small distinct heaps of white clouds; they are called fleecy clouds, from their resemblance to a collection of fleeces of wool, and are imperfectly represented on the general cut. They do not appear often, and are usually *fair-weather clouds*.

This form has none of the characteristics of the cumulus, and does not appear in the same stratum. It was probably called cumulus because its small masses are distinct, as are those of the ordinary cumulus. It occurs in the same stratum as cirro-stratus, and properly belongs to that modification. I retain the name inasmuch as the cloud is of some practical importance.

The cumulo-stratus is seldom seen in our climate, as it is represented in the cut. Stratus condensation *above*, and in connection with cumulus condensation, is not uncommon, but that precise form is rare.

This, too, is practically of no consequence, and I shall take no further notice of it.

Recapitulating, I give (in a tabular form) the three principal strata and their modifications, located with sufficient accuracy for illustration. The clouds which are found in an upper or lower portion of a stratum are so represented by the location of their names; those which appear at all heights in the stratum, with the names across. The elevation is the average one—although there is no limit to the cirrus above, except the absence of sufficient moisture. It was seen by Guy Lussac, and has been by other aeronauts, at an elevation of five miles, or more, when too delicate to be visible below.

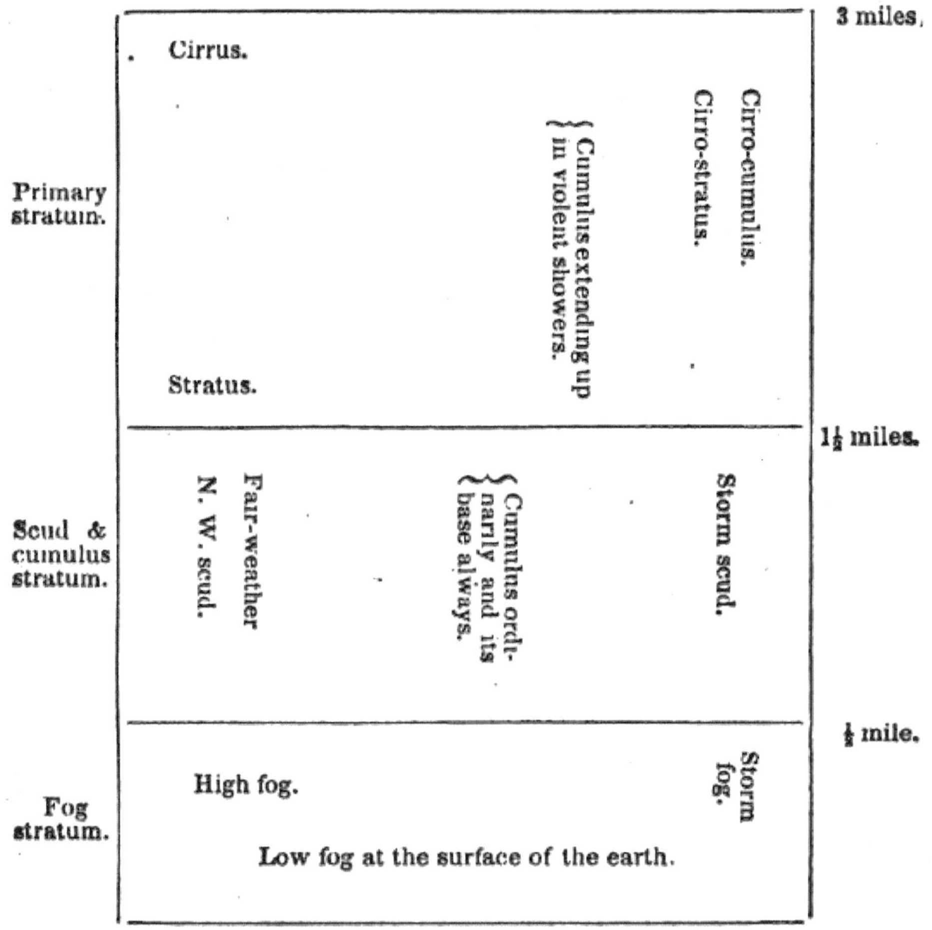

With the assistance of this table of elevations, and a careful observation, the reader can soon become familiar with the forms of clouds and their relative situations.

CHAPTER III.

Our rivers do not return from the North Atlantic—All storms and showers move from the westward to the eastward— Seeming clouds seen moving from the eastward to the westward are scud—They are incidents of the storm, and not a necessary part of it—The storm clouds are above them, moving to the eastward— Occasions when this may be seen —Admitted facts prove it—Investigations prove it—May be known from analogy—From the fact that there is an aerial current pursuing the same course in which the storms originate—Character of this current—Its influence upon our country—Importance of a knowledge of its origin, cause, and the reciprocal action between it and the earth—To this end necessary to go down "to the chambers of the South

Having thus taken a brief view of the different clouds, let us return to the inquiry, from what ocean, and by what machinery, *our "rivers return."*

Not wholly or mainly from the North Atlantic, although it lies adjacent to us, and they often *seem* to do so; for, first, all storms, showers, and clouds, which furnish, *independently*, any appreciable quantity of rain to the United States, and even adjacent to the Atlantic, or indeed to the Atlantic itself, come from a westerly point, and pass to the eastward. *This is a general, uniform, and invariable law, although there is in different places, and in the same place at different times, some variation in their direction; ranging in storms from W. by S. to S. S. W., and in showers between S. W. and N. W., to the opposite easterly points of the compass; the most general direction, east of the Alleghanies, being from W. S. W. to E. N. E.*

But do we not see, you inquire—at least those of us who live east of the Alleghanies—that when it rains, the wind is from the eastward; and that the *clouds* follow the wind from the east to the west? You do indeed, generally, in all considerable storms, observe that the wind blows from some easterly point, and that *seeming* clouds are blown by it to the westward; but what you see, and call clouds, are not the clouds which furnish the rain. Far above the seeming clouds you notice, directly over your head when it rains or snows, are the rain or snow clouds, dense and dark, passing to the eastward, how strong soever the wind may blow from the quarter to which they tend, or any other quarter, between you and them. What you see below them are *scud*. So the sailors call them, and so I have termed them. It is a "dictionary name," and a good one, expressive of a distinction between them and *clouds*. They are thin, and the sun shines through them, although with some difficulty, when the rain clouds above are absent or broken. *This east wind and the scud are not the storm, or essential parts of it.* Storms occasionally exist, particularly in April, without either. They are but *incidents, useful,* but not *necessary incidents,* as all surface winds are.

If you could see a section of the storm, you would see the rain cloud above, moving to the east, and the scud beneath running to the west, as indicated by the arrows in the cut on page 40. Opportunities frequently occur when these appearances may be seen. Storms are sometimes very long, a thousand miles, perhaps, from W. S. W. to E. N. E., and not more than one to three hundred miles wide from S. E. to N. W., and their sides, particularly the northern ones, regular, and without extensive partial condensation. Then the storm cloud above, moving to the eastward, and the scud running under to the westward, may be seen as in the cut.

So they may be seen before, at the commencement, and at the conclusion of easterly storms, in a majority of cases, and the reader is desired to notice them particularly as opportunities occur.

The term *running,* too, is a very expressive one, used by sailors as applicable to *scud.* For while the forming or formed storm clouds may be moving moderately along, at the rate of twelve to fifteen or twenty miles an hour, from about W. S. W. to E. N. E., the scud may be running under them in a different direction—opposite, or diagonal, or both—at the rate of twenty, fifty, sixty, and, in hurricanes, even ninety miles an hour. You have doubtless seen these scud running from N. E. to S. W., and without dropping any moisture, a day or sometimes two days, before the storm coming from the S. W. reached the place where you were; and then, sometimes the storm cloud slipped by to the southward, and the expected storm at that point proved "a dry northeaster." Sometimes the condensation, although sufficiently dense to influence and attract the surface atmosphere, and create an easterly wind and scud, does not become sufficiently dense to drop rain, and then, too, we have a dry northeaster, which may melt away or increase to a storm after it has passed over us. *I have never seen, except, perhaps, in a single instance, one of these masses of scud, however dense, which had not a rain (stratus) cloud above it, drop moisture enough to make the eaves run.* So you see it may be true, and if you will examine carefully, you may satisfy yourself that it is true, that the storms all move from a westerly point to the eastward, notwithstanding the wind under them is blowing, and the scud under them are running to the westward.

There are many other methods by which the reader may determine this matter himself. He may catch an opportunity for a view, when there is a break in the stratus cloud above, and the sun or moon, no longer obscured by the *storm cloud,* shines through the scud beneath. Then he may see they are moving in different directions. *The upper cloud, if there be any of it left, always to the eastward.*

Again, we may see the storm approach from the westward, as it often does, before the wind commences to blow, and the scud to run from the eastward; particularly snow storms in winter, and the gentle showers and storms of spring.

Again, thunder storms, we know, come from the westward, and apparently against an east wind. It is sometimes said they approach from the east, but it is a mistake. During thirty years attentive observation in different localities, I have never seen an instance. They sometimes *form* over us, or just east of us, or one may form at the east and another at the west, and as they *spread out in forming*, one may seem to be coming from the east, or there may be an easterly current, with dense flocculent scud at the under surface of the shower cloud running westward, but they finally pass off to the eastward, and never to the westward. It is possible that a *patch of scud* may become sufficiently *dense* and *electrified* to make a *shower*, but I have never observed one. Such an *apparent* instance may be found recorded in "Sillman's Journal," vol. xxxix. page 57. I have seen the scud assume a distinct cumulus form, but never to become sufficiently dense to make a thunder shower.

Thunder and lightning sometimes attend portions of regular storms in spring and autumn, but the thunder is always heard first in the west, and last in the east.

Again, there are admitted facts with which you are conversant, which prove this proposition. When it has been raining all day, and just at night the storm has nearly all passed over to the eastward, and the sun shines under the western edge of it, and "*sets clear*," as it is termed—you say that "*it will be clear the next day*." Why? Because the storm will not pass to the westward, covering the sun and continuing, how strong soever the wind may be from the east; and because it is passing, and will continue to pass off to the eastward, leaving the sky clear. *The easterly wind will stop as soon as the storm clouds have passed, and it will fall calm, or the wind will "come out" from the westward.*

So, too, when the clouds are dark in the west in the morning, and the sun rises clear, but "*goes into a cloud*," as it is expressed, you say that it will rain. And if the clouds are dense this generally proves true; because there is a storm or shower approaching from the west, and passing over to the east, the western edge of whose advance condensation has met the sun in his coming, and obscured him from your vision.

When, too, it has been storming, and lights up in the N. W. you say it will clear off; the N. W. wind will blow all the clouds away. It is, indeed, generally true that when it so lights up it is about to clear off; although it sometimes shuts down again, in consequence of the approach of another storm from the westward, following closely behind the one which is passing off. It is a great mistake, however, to suppose the N. W. wind blows away the clouds. Watch the smooth stratus rain cloud at its lower edge, where the clear sky is seen, and you will see that it is moving on steadily to the N. E., in obedience to the laws of its current, and will do so, even when its retreating edge has passed up to the zenith, and down to the S. E.

The storm uncovers us from the N. W. by the contraction of its width, *or* because it has a *southern lateral extension* and *dissolution*, and not by being blown away by the N. W. wind; although that wind, by its peculiar fair-weather clouds, may be, perhaps, observed beneath, ready to follow its retreating edge.

Again, when it has been clear all day, and the sun sets in a bank of cloud, you say—"*it will rain to-morrow, the sun did not set clear*," and unless that bank is a thunder cloud, merely, which will pass over or by you, with or without rain, before morning, it is generally true that it will. The bank will prove the eastern edge of an approaching storm.

From these generally admitted and understood facts, you may know that storms pass from the west to the east.

This proposition is also proved by all the investigations of storms, which have taken place since the settlement of this country. Storms of great severity attract particular attention, and are said to "back up" against the wind, because they are observed to commence storming first at the westward, although the wind is from the eastward. Doubtless you recollect many such instances recorded in the newspapers. No season occurs without such notices.

Many storms have been investigated by Mr. Redfield, for the purpose of sustaining his theory. Many others by Professor Espy, to sustain his. One by Professor Loomis, with great research and ability—and some by others, accounts of all which have been published; and every one yet investigated, north of the parallel of 30°, has been shown to pass from a westerly to an easterly point.

So, too, we may know it from analogy. The laws of nature are uniform. There is a great end to be accomplished, *viz.*: the distribution of forty inches of water, at regular intervals, over a large extent of country. The rivers are to return, and the clouds are to drop fatness, and seed time and harvest are not to cease. It is to be done and is done, by means of storms and showers, and pursuant to general laws, as immutable as the result. Most of these storms and showers, it has been found, and may be observed, move from the westward to the eastward. Then we may know, from analogy, that they do so in obedience to a general, uniform law; and so I might say with confidence, if our inquiry stopped here, it will ever be found by those who may hereafter examine them.

But, 2d. There is a current in the atmosphere, all over the continent north of the N. E. trades, but in great volume over the United States, east of the meridian of 105° W. from Greenwich—varying in different seasons, and upon different parallels, and flowing near the earth, when no surface wind interposes between them. In the vicinity of New York, the usual course of this current is from about W. S. W. to E. N. E. In the western and south-western portion of the United States, it is, doubtless, more southerly—varying somewhat according to the season—and in other sections varies in obedience to the general law of its origin, and progress.

I have observed its course in many places, between the parallels of 38° and 44° N. *This current comes from the South Atlantic Ocean.* It is our portion of the aerial current, which flows every where from the tropics toward the poles, to which I have already alluded in connection with the distribution of heat. *It brings to us the twenty inches of rain which we lose by the rivers, and by the westerly winds, which carry off a portion of the local moisture of evaporation, and its action precipitates the remaining portion of that moisture. It spreads out over the face of our country, with considerable, but not entire uniformity. All our great storms originate in it, and all our showers originate in or are induced and controlled by it.*

From the varied action, inherent or induced, of this current, most of our meteorological phenomena, whether of wet or dry, or cold or warm weather, result; and a thorough knowledge of its origin, cause, and the reciprocal action between it and the earth, is essential to a knowledge of the *"Philosophy of the Weather."*

Let us then go down to the "chambers of the south," to the inter-tropical regions, of which we have said something in connection with a notice of Southern Mexico, and see where, and how this great aerial current originates.

CHAPTER IV.

Between the parallels of 35° north latitude, and 35° south latitude—changing
its location within this limit at different seasons of the year—encircling the earth,
and covering about one-half of its area—we find the trade-wind region. In this
region are the simple and uniform arrangements, which extend every where, and
produce all the atmospheric phenomena. In the center of it we find that movable
belt of continual or daily rains, and comparative calms, particularly *near its center*,
about four hundred and fifty miles in width upon the Atlantic, and over Africa,
and the eastern portions of the Pacific, and something more over South America
and the West Indies, the western portion of the Pacific and the Indian Ocean, to
which we have already alluded. This belt of rains and calms follows the trades
and sun, in their transit north and south, from one tropic to the other—its width
and extension depending upon the volume of trade-winds existing on the sides of
it. Its southern edge, when the sun is at the southern solstice, extends to 7° south
in the Atlantic, to 10° south in the Indian Ocean, and still further, probably, over
South America: on this point I do not pretend to be accurate, for accuracy is not
essential. When the sun is at the northern solstice the southern edge is carried up
as far as 12° north, over the Atlantic, and still further over the northern portions
of South America, the West Indies, and Mexico. It travels, therefore, from south to
north, over from twenty to forty degrees of latitude. The presence of this belt of
rains over any given portion of the inter-tropics, gives that portion its rainy season,
and its absence, as it moves to the north, or the south, gives the portion from which
it has moved, its dry season. It passes in its transit twice each year over some
portions of the country, Bogota, for instance, and two corresponding rainy and
dry seasons result. Its presence, and character, and movements, are as fixed and
regular, over from twenty-five to forty degrees of the earth's surface, *and all around*

it, as the presence and movements of the sun over the same area.

At the northern edge of this movable belt of rain, and extending in some places, particularly in the Pacific Ocean, north about 20°, or about one thousand four hundred miles, and in other places a less distance, the N. E. trade winds prevail, blowing toward and into it from N. N. E., N. E., and E. N. E., averaging about N. E. At the south line of this belt of rains, extending south from twenty-five to thirty degrees, or from sixteen hundred to two thousand miles, the S. E. trades blow toward and into it, from the S. E., S. S. E., or E. S. E., averaging about S. E. Of course the northern limit of the N. E. trades travels north and south with the belt of rain, toward which it blows; and so the southern limit of the S. E. trades travel in like manner with the rainy belt, or rather, to speak with entire accuracy, the belt of rain moves with the trades, and the trades follow the verticality of the sun. The following diagrams exhibit approximately, and with sufficient accuracy for illustration, the situations of the rainy belt and the trades, when at their northern and southern limit, as well as the manner in which it must give certain localities two rainy seasons each year, in its transit north and south.

At the northern and southern limits of the trade-winds, and extending from them to the poles, are found the variable winds and irregular extra-tropical rains, all over the earth, which are shown by the shading on the maps. This line of extra-tropical rains descends to the south, following the retreating trades as they descend in our winter, and recedes north before the trades when they return in spring and summer, so that at the outer limit of the trades respectively, toward the poles, the line of extra-tropical rains will be found, receding or following that limit, as the trades pass up and down with the sun. From the north pole to the northern limit of the N. E. trade-winds, wherever found, whether at 38° north latitude, as in some places in summer when the sun is at the tropic of Cancer; or whether at 20° to 30° north latitude, as in our winter, when the sun is at the tropic of Capricorn; the extra-tropical rains prevail. A state of things precisely similar exists between the south pole and the southern limit of the S. E. trades. Between this northern limit of the N. E. trades and the northern line of the inter-tropical belt of rains, wherever situated (with two exceptions, to which we have alluded and shall allude again), there is, for the time being, a dry season; and a like dry season between the southern line of the belt of rains and the southern limit of the S. E. trades. We have, therefore, extending around the earth, a belt of daily tropical rains, near the center,—two belts of drought which are mainly trade-wind surfaces, one on each side of the central rainy belt,—extending to the outward limits of the trades and the line of extra-tropical rains; and these rainy and dry belts, moving up and down after the sun, a distance of from twenty to forty degrees of latitude, each year.

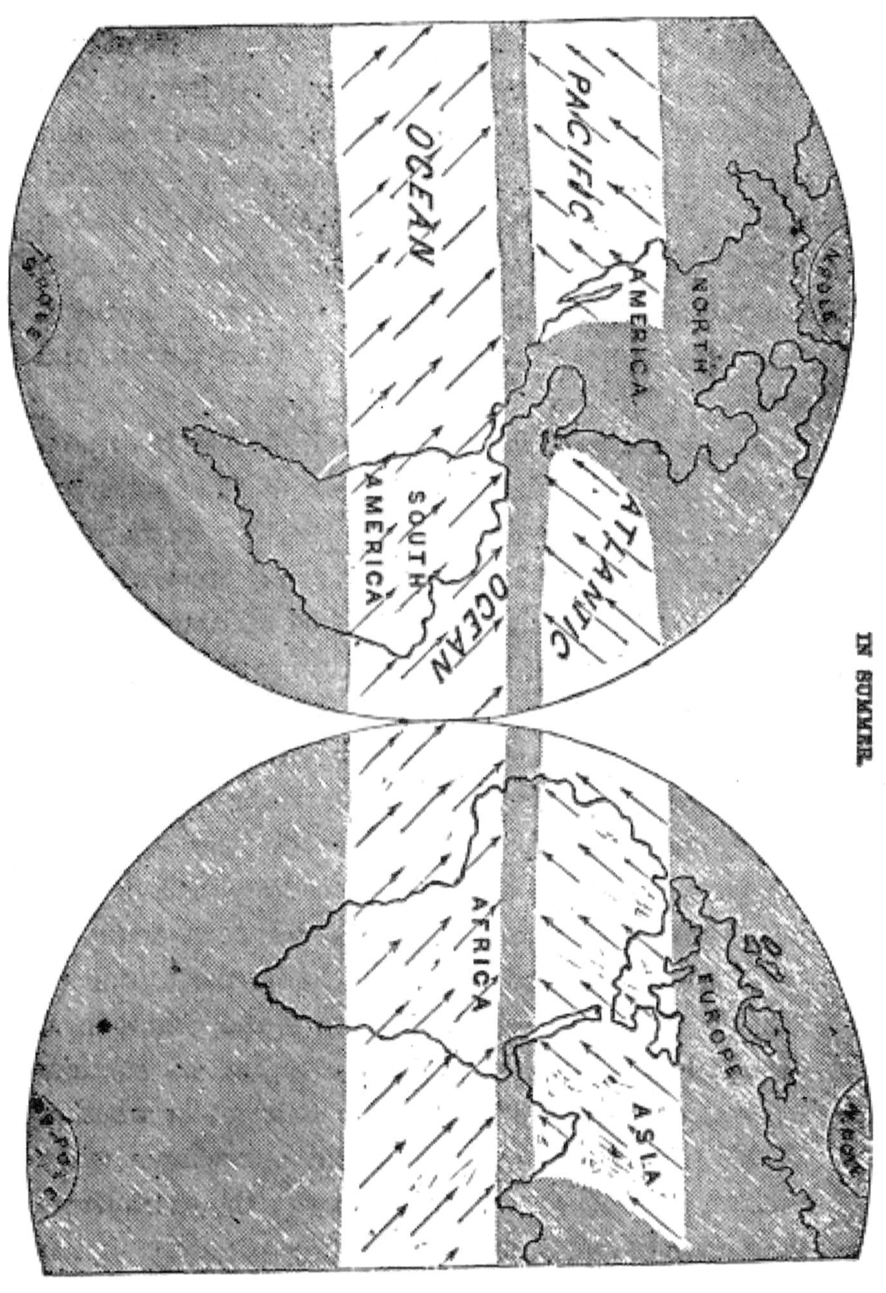

FIG. 10.

IN SUMMER.

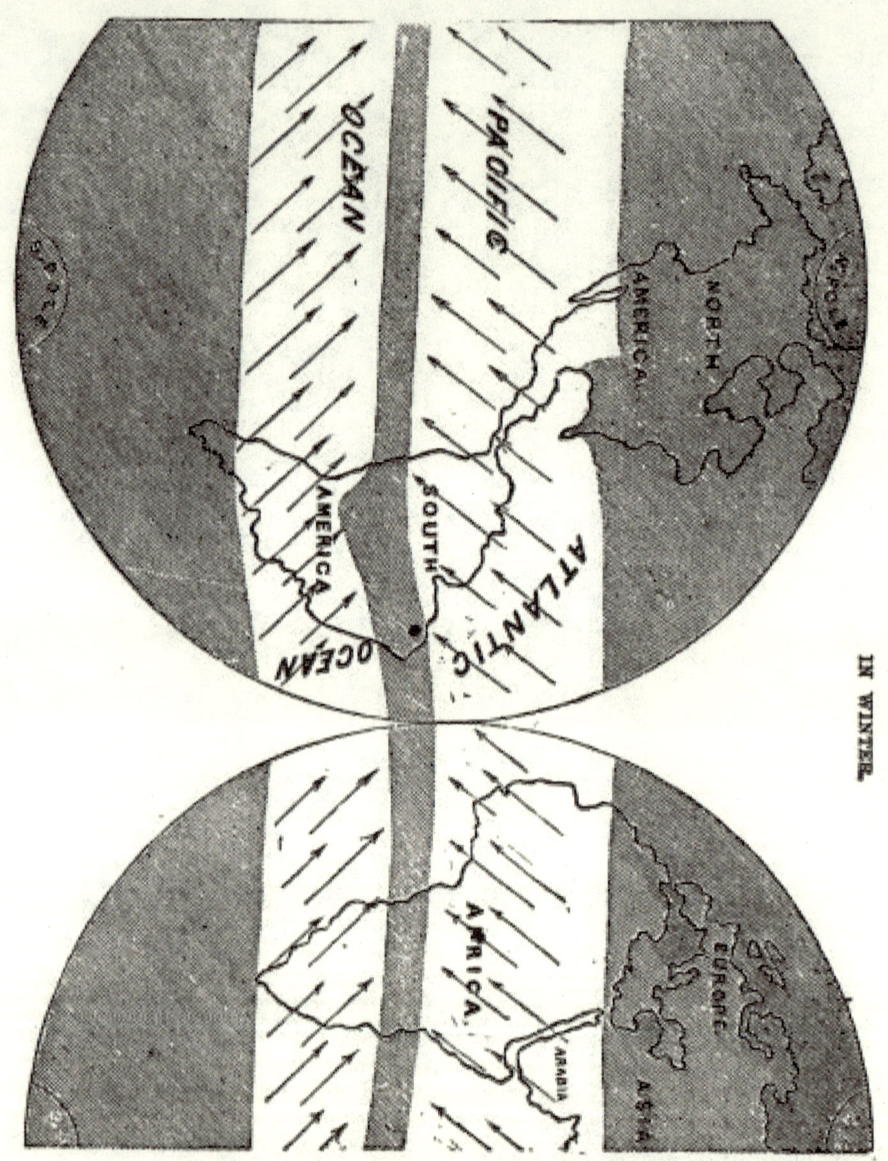

FIG. 11.

IN WINTER.

Such are the *main* phenomena, *at the surface*, in the trade-wind region. Ascending a step higher in the atmosphere, we find, above the surface-trades, a counter-trade, running, not in the opposite direction, but at right angles, or nearly so. The counter-trade which issues from the northern side of the rainy belt, running to the N. W. or W. N. W., and the counter trade which issues from the southern side, running to the S. W. or W. S. W., varying, as the trades do in direction in different localities. These counter-trades are continuations of the surface trades, which, ascending in their course, have threaded their way through the opposite trade in the rainy belt, and are continuing on at the same angle, and in the same direction at which they blew upon the surface, and in obedience to the same law. This is apparent from several considerations.

1st. They issue at the same angle, and over the top of the surface trades. In the West Indies and elsewhere, this has been ascertained and proved by the course of the storms, and the rotation of their surface winds, and observation.

2d. We can not suppose the N. E. trade to be reflected, and turn back over itself at a right angle. That would be impossible, even if there were a wall of solid material there for it to blow against. Air is a peculiar fluid, and it stratifies with astonishing ease. He who supposes that a current of air put in motion can be turned aside by another current, or by the atmosphere at rest, or can be made to mingle, is mistaken. It will stratify, and force itself onward through the adjacent and opposing atmosphere, and in a right line. I have observed some remarkable instances of this character.

3d. The cause which operates to produce the surface trades, still operates upon the current to carry it over into the other hemisphere; a counter-trade, as we shall see. It is impossible, therefore, to believe that the surface-trades as they arrive at the belt of rains and calms, turn at a right angle, or at any angle, and return: and impossible to doubt that they pass through each other in this belt, and out at the opposite side, as upper currents, at the same angle at which they entered. Of course the N. E. trade of the Atlantic becomes the N. E. counter-trade of South America, carrying their storms in a S. W. direction, and the S. E. trade of the Atlantic the S. E. counter-trade of the West Indies, carrying all their storms in a N. W. direction; and what is true of them is true of the trade winds *every where, all over the globe, over the land and over the sea.*

Doubtless here some one will say, our upper current is a S. W. current. True, the S. E. trade which enters the belt of rains, and issues out on the north, a S. E. upper current or counter-trade, keeps that course until it arrives at the northern limit of the surface trade, when, in *obedience to another law*, which we shall notice, it gradually *decends near the surface, curves to the eastward*, and becomes *the S. W. current which passes over us*. And so we have the S. E. trade-wind of the South Atlantic, with its moisture, warmth, electricity, and polarity, over, and perhaps sometimes around us, dropping the electric rain which makes glad our fields; giving us, when not prevented by other conditions, the balmy air of spring, the Indian summer of autumn, and the mild mitigating changes of winter; and thus, *our rivers, which run into the sea, return to us again.*

But let us go back to the trade-wind region—the region of regularity and uniformity—and examine somewhat more attentively its features, that we may more fully understand the character of this counter-trade.

Here are 60° at least of the 180° of the earth's surface, and at its largest diameter, covered in the course of the year, and of their travels, by the trade-winds at the surface, the counter-trades above, and the belt of rains and comparative calms, formed by the action of the opposite trades, as they thread their way through each other, to assume the relation of counter-trades. Truly the magnitude, simplicity, and regularity of this machinery are most wonderful.

There are, however, some *apparent* anomalies which deserve attention. Here are most distinctly marked the *rainy* and *dry seasons*, existing side by side. Here are the *rainless portions* of the earth, already but briefly alluded to; here the *monsoons*, and another peculiarity, *viz.*: the *gathering of the counter-trades* upon the western sides of the two great oceans, into two *aerial currents of greater volume, analogous* somewhat to the two *gulf streams* of those oceans. Let us examine these anomalies.

The rainy and dry seasons depend, as we have seen, upon the transit north and south of the rainy belt, or belt of comparative calms. Wherever this belt may happen on any given day to be situated, each side of it the trades prevail, it is dry, the earth is parched, and vegetation withers. These changes are graphically described by Humboldt in his "Views of Nature," as they occur on the northern portions of South America, as follows: "When, beneath the vertical rays of the bright and cloudless sun of the tropics, the parched sward crumbles into dust, then the indurated soil cracks and bursts, as if rent asunder by some mighty earthquake. The hot and dusty earth forms a cloudy vail, which shrouds the heavens from view, and increases the stifling oppression of the atmosphere; while the east wind (*i. e.* trade-wind), when it blows over the long heated soil, instead of cooling, adds to the burning glow.

"Gradually, too, the pools of water, which had been protected from evaporation by the now seared foliage of the fan-palm, disappear. As in the icy north animals become torpid from cold, so here the crocodile and the boa-constrictor lie wrapped in unbroken sleep, deeply buried in the dried soil. Every where the drought announces death, yet every where the thirsty wanderer is deluded by the phantom of a moving, undulating, watery surface, created by the deceptive play of the reflected rays of light (the mirage). A narrow stratum separates the ground from the distant palm-trees, which seem to hover aloft, owing to the contact of currents of air having different degrees of heat, and therefore of density. Shrouded in dark clouds of dust, and tortured by hunger and burning thirst, oxen and horses scour the plain, the one belowing dismally, the other with outstretched necks snuffing the wind, in the endeavor to detect, by the moisture in the air, the vicinity of some pool of water not yet wholly evaporated.

"Even if the burning heat of day be succeeded by the cool freshness of the night, here always of equal length, the wearied ox and horse enjoy no repose. Huge bats now attack the animals during sleep, and vampyre-like suck their blood; or, fastening on their backs, raise festering wounds, in which mosquitos, hippobosces, and a host of other stinging insects, burrow and nestle. Such is the miserable existence of these poor animals, when the heat of the sun has absorbed the waters from the surface of the earth.

"When, after a long drought, the genial season of rain arrives, the scene suddenly changes. The deep azure of the hitherto cloudless sky assumes a lighter hue.

Scarcely can the dark space in the constellation of the Southern Cross be distinguished at night. The mild phosphorescence of the Magellanic clouds fades away. Even the vertical stars of the constellations Aquila and Ophiuchus, shine with a flickering and less planetary light. Like some distant mountain, a single cloud is seen rising perpendicularly on the southern horizon. Misty vapors collect and gradually overspread the heavens, while distant thunder proclaims the approach of the vivifying rain. Scarcely is the surface of the earth moistened, before the teeming steppe becomes covered with Killingiæ, with the many-panicled Paspalum, and a variety of grasses. Excited by the power of light, the herbaceous Mimosa unfolds its dormant, drooping leaves, hailing, as it were, the rising sun in chorus with the matin song of the birds, and the opening flowers of aquatics. Horses and oxen, buoyant with life and enjoyment, roam over and crop the plains. The luxuriant grass hides the beautiful and spotted jaguar, who, lurking in safe concealment, and carefully measuring the extent of the leap, darts, like the Asiatic tiger, with a cat-like bound on his passing prey."

Such is Humboldt's description of the dry season on the Orinoco, and the return of the belt of rains from the south.

Again, within this trade-wind region are the *rainless countries*. These are portions of the earth which the equatorial rainy belt does not ascend far enough north in summer to cover, nor does the southern edge of the extra-tropical regular rains descend, in winter, far enough south to cover them, and where, of course, rain seldom, if ever, falls. Such are the central parts of the Desert of Sahara, Egypt, Arabia, portions of Affghanistan, Beloochistan, and the western parts of Hindoostan, to the north of the inter-tropical belt, and a similar state of things exists south of the equator in parts of South America, Africa, and New Holland, although upon a comparatively small surface.

Again, another anomaly is the gathering of the trade winds into greater volumes, on the westerly side of the great oceans, and the consequent carrying of the equatorial rainy belt up to the region of extra-tropical rains, on the eastern side of the great continents of Asia and North America, and the peculiar liability of these aerial gulfs to hurricanes and typhoons. Such an aerial gulf gathers over the Caribbean Sea, and the West Indies. Passing across the Gulf of Mexico, it enters over Texas, and Louisiana, and the other southern states; its western edge passing north in autumn and winter, on the eastern side of the highlands of Western Texas, New Mexico, and the Great Desert; curving, as all counter-trades do, to the eastward as soon as it passes the limit of the N. E. trades, and spreading out over our favored country, leaving the evidence of its pathway in the greater quantities of rain, which fall annually upon its surface. This gathering deprives a portion of the Atlantic, north of the tropics, of its share of the counter trade, and there, as every where, where the volume of counter-trade is small, storms and gales are infrequent, and of less force, and comparative calms prevail. That portion of the Atlantic has long been known as "the horse latitudes," a name given to it by our Yankee sailors, because, there, in former times, the old-fashioned, low-decked, flat-bottomed, horse-carrying craft of New England, bound for the West Indies, often floundered about in the calms and baffling winds, until their animals perished for want of water, and were thrown overboard. Lieutenant Maury, in his most praiseworthy and exceedingly useful investigation of "The Winds and Currents of the Ocean," has defined the situation of these calms and baffling winds at different seasons—for they move up and down, of course, with the motion of the whole machinery—and enabled navigators to avoid them, by running *east* before they attempt to make *southing*; and very materially shortened the voyages to the equator.

A like gathering, in volume, of the S. E. trade, on the western side of the Pacific, enters over Asia, and covers China and Malaysia, extending, in its western course, nearly as far as the western edge of Hindoostan. In this concentrated volume of counter-trade, and owing to its concentrated action, form and float the typhoons of the China Sea, and of the Bay of Bengal; and to this anomalous aerial gulf stream, the S. E. portions of Asia, from the western desert of Hindoostan, to the eastern portion of China, north of the rainy belt, owe their great supply of moisture and fertility, and their peculiar climate. The western line of this volume of counter-trade is marked by the eastern portion of the rainless region of Beloochistan, and the north-western deserts of India, as the western edge of our concentrated volume of counter-trade, is marked by the arid plains of northern Mexico, western Texas, and New Mexico. On the south of the equatorial rainy belt, there is no corresponding aerial gulf of equal volume, as there is no corresponding gulf stream of equal magnitude. On the western side of the Indian Ocean we find a gathering of the N. E. trades from the Bay of Bengal and the Indian Ocean, in which form and travel the hurricanes which prevail—traveling to the southward and westward—about the Isle of France or Mauritius; and the lagullus oceanic current, which runs down to the S. W. toward the Cape of Good Hope. But the extension of South America to the eastward, under, or just south of the N. E. trades, does not permit the formation of such a concentrated volume on the western side of the Atlantic, nor is the strength or regularity of the N. E. trades, on that ocean, equal to those of the S. E.

Nor is the magnetic intensity on the eastern and middle portions of the Pacific, sufficient to produce such a concentration, in large volume, there. The trades over that ocean, therefore, curve without concentration, except a partial one, over the western groups of Polynesia, which the Asiatic line of magnetic intensity approaches and where hurricanes are sometimes found, until we arrive near the eastern line of magnetic intensity, on the eastern side of Asia. We shall, hereafter, have occasion to follow the anomalous concentrated volumes of the S. E. counter-trade, of the northern tropic, on the western side of the great oceans, in explanation of some of the phenomena which we find north of the trade-wind region. Suffice it here to add, that if it were not for the concentration of these counter-trades, on the western side of the great oceans, the rainless region between the parallels of 20° and 30° would encircle the earth; and China and the Eastern United States would have a distinctly marked rainy and dry season, as have California, the Barbary States, Syria, Persia, and other countries which lie north of the rainless region, within the summer range of the N. E. trades, but also within the winter descending range of the belt of extra-tropical rains.

Another anomaly which we find in the trade-wind region, is the monsoon.

There are several of them, but they are found, in the greatest strength and regularity, in the Indian Ocean. Another, defined by the investigations of Maury, is found on the west coast of Africa, extending out over the Atlantic. Another prevails on the western coast of South and Central America. The etesian winds of the Mediterranean are but the N. E. trades, whose northern limit is carried up in summer, by the transit of the connected machinery, to the north, over that sea. The N. E. and S. E. monsoons, so called, of the Indian Ocean, are but the regular trades, blowing when the belt of rains is absent, as they do all over the globe. The N. W. monsoon, south of the equator, in the vicinity of New Holland; the S. W. monsoon which blows from the Arabian Sea, in upon Hindoostan; the S. W. monsoon of the Atlantic, south of the Cape De Verde Islands; and the variable west monsoon winds of the west coast of Southern and Central America, and Southern Mexico (known under several different names, but chiefly by that of Tapayaguas), are all that deserve attention as such.

At first sight they appear to be anomalies, but the facts declare their character with perfect certainty. First, they are not continuous, like the trades, but *prevailing* winds, and are *storm winds; they always blow toward a region, or portion of the ocean, covered at the time by clouds and falling weather.*

Second, they do not blow upon, or toward, heated surfaces of land or water — *i. e.*, toward the dry and parched surfaces, where the dry season prevails, or from adjoining cold waters on to warm surfaces, but toward the land or water *situated under the rainy belt.* They are therefore incident storm winds, (as our easterly winds are incident storm winds) of the rain clouds of the tropics. They blow in upon the land, under the belt of rains, while that belt with its daily cloud, and inducing electric action, is over it, and follow that belt in its transit north and south. They blow from the warm south polar current of the Atlantic, which flows N. W. from the coast of Africa, toward the inshore north polar current, which is there flowing south, but under the belt of rains. In the Indian Ocean they blow from the center of that ocean, and the Arabian Sea, toward the belt which hangs over Hindoostan, from the S. W.; and when the rainy belt travels south they still blow toward, and under it, from the Indian Ocean, but of course from the N. W. The heated character of the waters of the Indian Ocean and Arabian Sea, which receive no polar currents, but heated waters from the Persian Gulf, and from rivers which flow into the Bay of Bengal over the heated plains of a tropical country, explain this. So, too, the monsoon of the Atlantic Ocean, does not blow north of the Cape De Verde Islands, — where the heated surface of Sahara, burning with the rays of a vertical sun, has a temperature sometimes ranging from one hundred and forty to one hundred and sixty degrees — but remains under the rainy belt, drawn from the heated waters which flow up from the South Atlantic, and travels north as the rainy belt travels north in summer, and south to the Gulf of Guinea, as that travels south in winter. The same is true of the Pacific monsoon, the Tapayaguas, the least marked of all, which blows in during the rainy season upon the west coast of Southern Mexico, and of Southern and Central America. They are all incident rain or storm winds, blowing in upon the land, or on to a colder surface of different polarity, *during the rainy season;* and if it were possible to catch one of our north-easters, in its passage over our country to the eastward, and anchor it to the Alleghanies, "paying out" so to have it reach in part over the Atlantic, and keep it there in operation six months, we should have a continual easterly wind under it; a *monsoon* more strongly marked than the monsoons of the Indian, or Atlantic Oceans. *The received theory in relation to them is a fallacy.*

Recapitulating, then, all the phenomena, we have,—*Surface-trades*, blowing toward the center, passing through each other, and continuing on as upper or counter-trades; a *belt of rains*, with calms near the center, formed by the trades where they meet and pass through each other, which travels with them north and south following the sun; *two belts of drought*, following the belt of rains and the trades, and followed by the *extra*-tropical line of rains, as it travels with the trades and the rainy belt, leaving a part of the earth which the equatorial rainy belt does not travel far enough north, nor the extra-tropical line of rains far enough south to cover, and which is consequently a *rainless region*; *the monsoons*, which are but incidents of the rainy belt, and the *gathered volumes* of counter-trade, on the west of the two great oceans, which usurp the place of the N. E. trades, carrying the rainy belt up to the region of extra-tropical rain, and preventing the rainless region from encircling the earth.

Upon *what cause* do these great central phenomena, so vast, so regular, so wonderful, depend? What is the *motive power* of this connected atmospheric machinery, whose action and influence extend over the entire globe?

"*Heat, heat,*" say the text books, the Professors, the votaries of meteorology. "All these phenomena are owing to the heat of the sun. It heats the ocean and the earth—the air is thereby heated and rises, the cold air rushes in from below, then the ascended current rolls off each way at the top toward the pole, acquiring a westerly motion from the rotation of the earth, slipping away from under it, and a different, *viz.*: an easterly motion, after reaching the latitude of 30°, from the *same rotation*; and all the winds and disturbances of the atmosphere are produced in the same way. They are produced by the action of heated surfaces upon the adjacent atmosphere."

This is the great theory of meteorologists, by which they attempt to account for the various atmospherical disturbances, of both tropical and extra-tropical regions.

The whole theory is a fallacy—it will not stand the test of a careful examination. The bases of the theory, which are assumed to be facts, are not so. The agent has not the power claimed for it. A heated surface, alone, never caused any considerable ascending current, or if it did, never produced a mile of wind. I repeat it, the theory and all incidental ones—the thousand explanatory and modifying theories, and hypotheses—*the whole system*—is without foundation in fact, and will not bear a critical examination.

Let us see if this language is stronger than the facts will warrant.

The theory assumes that both the land and water, under this central belt, where the air is supposed to be rising are *materially hotter* than the land and ocean are on *either side of it*. Now, how much hotter are the air and the land under the belt of rains and calms, upon Hindoostan, or Africa, or South America, where the former is supposed to be acquiring heat and expansion so rapidly, and to be ascending, than under, and in the dry belts on either side? None; it is cooler by the thermometer—*much cooler.*

The central belt of rains in midsummer over Africa, extends up as far as 17° north latitude, and perhaps further. North of this line over the whole surface of the desert, the Barbary States, a part of the Mediterranean, and some portion of Italy, the dry season extends, and from the entire surface the N. E. trade blow into the central belt.[1] Over the desert they all pass. Now this desert is a sea of sand, under a vertical sun, intensely heated, blistering the skin with which it comes in contact, and often acquiring a temperature of 150° to 160° of Fahrenheit. Under the central belt of rains neither the earth nor air exceed the temperature of 84°. And yet the hot air of the desert does not ascend, but blows into this cooler central belt; and when it is felt as it blows off the western coast by the mariner, or even in Guinea, when the belt of rains has gone south in winter, as it often is as the *harmattan,* it is suffocating and intolerable. There, then, not only is it untrue, that the land and the air over it under the rainy belt are hotter, but it is true that intensely heated air blows horizontally from the Desert of Sahara. Nay, as it will appear in the sequel, this hottest of all surfaces not only can not have a vortex, but it can not induce a monsoon, and scarcely a sea breeze. The same is true in a great degree of the surface, and the air over it, on either side of the supposed vortex of the rainy belt upon South America. See the description of Humboldt, already given, where the thermometer stood as high as 115° of Fahrenheit in the shade, while the N. E. winds, the regular trades, were blowing over the land. And it is equally true of Arabia, and indeed of every portion of the earth. There is not a spot upon the globe where the land and the air are cooler *by the side* of the central belt of rains, than *under it. And the opposite is true every where upon the land.*

How much hotter is the ocean and air under this supposed vortex? But little hotter than they are on the side where the sun is not vertical, *and none on the other.* Let us be a little more particular. The temperature of the Atlantic under the belt of rains in our winter, and on the south of the belt at the latitude of 3° south, and down to 9° or more south, is 82°. The air may range a degree, or possibly two, higher than the water at either point. On the north this difference is from nothing at the meeting of the trades and belt of rains, to about 4° at their northern limit. This is too *trifling* to be worth one moment's consideration. It is less, far less than the difference between the water and air of the Gulf Stream which runs along our coast, and the adjoining waters and air over them. While on the south side of the belt of rains the *difference is actually against the theory*—and the same state of things is reversed in summer, when the sun is vertical at the north.

From the log of an intelligent shipmaster, found in the wind and current charts of Lieutenant Maury, I abridge the following, which will illustrate this. Captain Young in February, found the N. E. trades at about 17° north latitude, with the water at 75° and air at 76°, trade-wind N. E.

---

[1] See the diagram for summer at page 35.

	At	12° 16'	the water was	75°	the air	76°	wind	N. E.
Feb.	22nd.	9° 49'	"	76½°	"	77°	"	N. E.
"	23d.	7° 13'	"	78°	"	78°	"	N. E.
"	24th.	no obs.	"	79½°	"	79°	"	N. E., E. S. E. rain.
"	25th.	3° 10'	"	81°	"	83°	"	E. S. E. rain.
"	26th.	no obs.	"	82°	"	82°	"	S. E. to E. S. E. hazy, rain & sqs.
"	27th.	2° 24'	"	82°	"	82°	"	calm, with rain.
"	28th.	no obs.	"	82°	"	82°	"	calm rain.
March	1st.	0° 29'	"	82°	"	82°	"	E. S. E. sqs. rain.
"	2nd.	1° 27' S. L.	"	82°	"	82°	"	S. E. sqs. rain.
"	3d.	2° 44'	"	82°	"	83°	"	S. E. & S. S. E. weather settled.
"	4th.	4° 17'	"	82°	"	83°	"	S. S. E. & S. E. fair weather.
"	5th.	6° 08'	"	82°	"	84°	"	S. E. fair wthr.
"	6th.	8° 08'	"	82°	"	84°	"	S. E. & E. S. E. fair weather.

Here the air was seven degrees colder at the extreme limit of the N. E. trades than in the *center* of the belt of rains, as it is, usually, in mid-winter, but not in summer. On the other hand, *after he left the region of calms and rains*, where the water and air stood with almost entire uniformity at 82°, on the 3d of March, and for three days thereafter, during which he was in the S. E. trades with fair weather, the water was the same as under the supposed vortex, *viz.*, 82°, *and the air rose to 83° and 84°! This is demonstration.*

I also take from a letter of Lieutenant Walsh to Lieutenant Maury, relative to the cruise of the "Taney" the following, showing the warmth of the Gulf Stream compared with the adjacent ocean.

"We first crossed the Gulf Stream on the 31st of October; we struck it in latitude 37° 22', longitude 71° 26' as indicated by the temperature of the water, which was as follows:

8	A.M.	water at	surface	66°
9	"	"	"	73°
10	"	"	"	76°
11	"	"	"	77°

77° was the highest temperature found in crossing at this time.

Re-crossing it in May, in latitude 35° 30', longitude 72° 35', he found the water as follows:

8	A.M.	water at	surface	71° 8'
9	"	"	"	73°
10	"	"	"	75° 5'
11	"	"	"	78° 5'
12	M.	"	"	78° 5'

79° being the highest temperature found."

The average difference between the temperature of the water of the Gulf Stream and the adjoining ocean, at the line of division, is about ten degrees, increasing to more than twenty on approaching the coast, and within one hundred miles—a far greater difference than is ever found on the winter side of the inter-tropical rainy belt.

It is not only not so, then, that the surface of the ocean is materially warmer under the belt of rains than the adjoining surface under the trades, especially on the summer side, but if it were so, the trades would not be created thereby, any more than upon the Gulf Stream. And the opposite is true of the land where the line of calms, and rains, and drought meet, all around the globe. The fact assumed is therefore untrue. The hottest surfaces, even at the rainless portion, where there is no vortex, no storm, and no wind but the continual uniform N. E. horizontal trade-wind, *never* created, by reason of the heat alone, a mile of wind, a storm or shower.

But, again, the belt of calms, where the air is supposed to rise and create a suction which draws the trades on either side a distance of from one thousand to two thousand miles, an average of three thousand miles in all, at least, is not itself, on an average, over five hundred miles in breadth from north to south. What a wonder of meteorology is here!

With a breadth of five hundred miles, the rising of the atmosphere is supposed to be so rapid and of such immense volume that it draws the surface atmosphere, one thousand to fifteen hundred miles on one side and two thousand on the other, with a uniform steady velocity of twenty miles per hour. Is this vast suction found by the unlucky mariner who may be drawn within the vortex? *Not at all.* He finds no rapid suction there, but *horizontal currents*, not steady, indeed, like the trades, and sometimes calms *at the center*, but still the *currents are there*, and, *except near the center, there as squalls, showers, and baffling winds and as monsoons.*

Again, is there at the mouth of this vortex, or as you approach it, an increased rapidity in the trade corresponding to the magnitude of its influence? Does the trade become a hurricane as it approaches the spot where it is to supply the place of that which has suddenly "expanded by heat, and been forced to rise, boil over, and run off at the top in turn?" Not at all. It blows gently, even up to the very line of the rainy belt, and becomes squally and baffling, falls gradually calm near the center, or changes to a monsoon.

But, again, the belt of rains is so far from being a belt of calms strictly, that its monsoons in the Indian, Atlantic, and Pacific Oceans, at times, extend hundreds of miles out over the ocean. That of the Atlantic, triangular, with its base resting on Africa, according to Lieutenant Maury, extends sometimes almost to the coast of South America, a distance of one thousand miles, and thus under the supposed ascending vortex. Where is the great uprising suction during the prevalence of this extensive surface horizontal monsoon beneath it? Manifestly it does not exist. Nay, that monsoon is blowing from the warm current which sets up from the Cape of Good Hope toward the Caribbean Sea, and over the cold north polar current, which runs down between the continent and the Cape de Verdes. Equally untrue is the presumption that the air rises over heated portions of the earth elsewhere, and by reason of such heating. *Perpendicular currents of the atmosphere are rarely seen, never extensive, or attaining any considerable altitude.* I have watched for them thirty years. I have seen currents of air ascend, with their moisture condensing as they ascended, and unite with the under surface of a highly electrified cloud—the advance condensation of a thunder shower—but that cloud was moving horizontally at a distance of from one to two thousand feet above the surface of the earth, and did not rise. I have seen patches of scud rising from the surface during the intervals of a showery and highly electrified storm, toward, and uniting with, the clouds above, when very low, as I have seen them approach and unite horizontally; and doubtless there is a tendency upwards of the wind, created and attracted by the summer shower, as may be seen in the ascending dust before the rain, but I have never been able to detect an ascending current, except as induced and attracted by a cloud above moving horizontally, in the hottest day or dryest time. None of the clouds of our climate, even when the earth is heated and parched by a two months' unbroken drought, can be detected rising above the strata in which they form. I have watched the cumuli at such periods when they filled the air, and can assert that they never rise. The atmosphere moves, invariably, in horizontal strata, and the whole theory of ascending currents is fallacious.

But let us look still further at the tropical currents. The true harmattan of north-western Africa (for the term is sometimes misapplied), hot and blistering, generated upon the sand of the desert—why does it blow from Sahara horizontally, on or over cooler surfaces, following the belt of rains as a N. E. trade? Why does it not ascend? The sirocco of north Sahara, the kamsin or chamsin of eastern Sahara, and the simoon of Arabia, which blow hot and suffocating from those deserts—why do they blow *from* heated surfaces and *horizontally over* cooler ones? Why do they not ascend? Arabia is surrounded on three sides by seas and gulfs, from which evaporation is rapid. Her interior deserts are extensive and intensely hot—why are they rainless? Why do they not have a *vortex*, a *monsoon*, or even a *shower*? Because there is no such law or action as this theory supposes. Those winds blow horizontally in obedience to other laws, and under the control of other and more powerful agents. But further still, what heating and ascending process is it that makes the variable winds north of the tropics? that brings in the warm air and fog of the Gulf Stream upon our *snow-clad coast*, in mid-winter, to increase the January thaw? Nay, what heating process is it that disturbs the calms of the polar regions with fresh breezes and gales, sometimes of the force of 6, when the *sun does not shine*, the thermometer is from 20° to 40° below zero, the *earth and sea one frozen surface*, and the hardy explorer dressed in furs, barely lives in his cabin covered by an embankment of snow, and heated by a stove?

Gentlemen, meteorologists, it will not do. The theory is unsound; the assumed facts do not exist. The whole universe has not an agent, organic or inorganic, which can play such absurd and inconsistent pranks in the face of its Creator, as your various and complicated theories assign to caloric.

Away with the theory and all its incidental and complicated and mystified hypotheses, they rest like a pall upon the science;—away with the whole system, and let us seek some agent whose *power* and *adaptation* correspond with the *extent*, and *simplicity*, and *magnificence* of the phenomena, and, in some degree, with the *power* and *wisdom of their Author*.

CHAPTER V.

The agent, magnetism—Its character and currents—Oxygen magnetic—Precipitation at the belt of rains occasioned by depolarization—Storms originate in this central belt, and move toward the

One, and the principal end attained by the power of the agent, is the gathering of a volume of atmosphere from, or near, the *surface* of the land and sea, so as to ensure its possession of all the moisture of evaporation which rises from the locality, and the highest degree of temperature, and from a space ranging from one to two thousand miles in width, in one hemisphere, and to carry it over into the other. Not over the top, or upon the top, of the whole mass of atmosphere situated in the opposite hemisphere—*out of reach of all influences from the earth*—but through it, and curving gradually down near to, and within influential distance of the surface of the earth, soon after it passes the outward limit of its fellow trade; and to continue the current onward, leaving portions of it and its heat and moisture on the way, but taking a considerable volume up and around the magnetic poles—it being impossible for the entire volume to be thus carried around the poles in consequence of the diminished circumference of the earth. To this end it is obvious it must possess *polarity.*

Another end to be attained is to combine the moisture of evaporation with the air, so that the cold atmosphere through which, or the earth over which it passes, may not be *continually condensing its moisture,* and thereby *enveloping the earth in a perpetual mist*; but so that it may part with it at *intervals,* making *cloudy* and *clear days*; and part with it in *portions,* so that a *regular* and *necessary supply* may be furnished to the *entire hemisphere,* even up to the geographical poles. Is there such an agent? There is, precisely and perfectly adapted to the ends to be attained, ever there and ever active, and that agent is *magnetism.*

The earth is a magnet. It has its magnetic poles, and they are distinct from its geographical ones; and there are two in each hemisphere. They are situated from 17° to 19° distant from the geographical poles; and ours is not far from longitude 97° W. from Greenwich, and 71° north latitude. Navigators have gone north and north-west of it, and found its situation by the declination of the needle. From these poles, lines of magnetic intensity extend to the opposite and corresponding pole of the other hemisphere, and upon or near those lines the needle points north without variation; and toward these lines of no variation the needle every where, on either side declines. The foregoing diagram shows the situation of our magnetic pole and line of no variation, the dip of the needle by the arrows, and the magnetic equator.

FIG. 12.

Recent discoveries have shown that the magnetic force is exerted in lines and currents; that such currents, as physical lines of force, surround magnets, and currents of electricity. Doubtless such lines of force exist around the earth and the magnetic poles. There are also *longitudinal* lines of force existing and active, between the poles, and extending from one side of the center to the other, occupying nearly one third of the magnet. If you take a large needle thoroughly magnetized, place it upon paper and drop filings of iron upon it, they will become arranged about it in circular and perpendicular, and also in *longitudinal lines*, conforming to the currents.

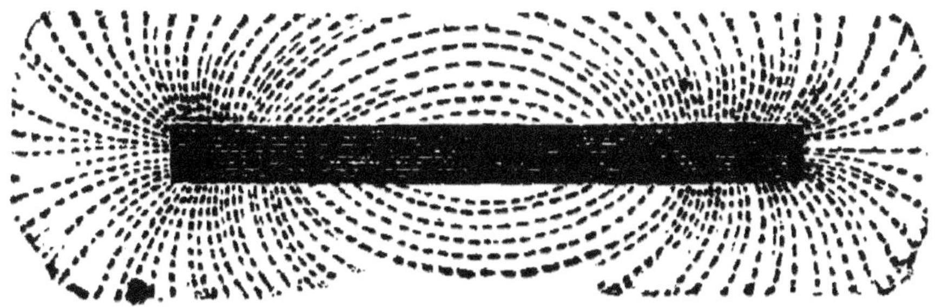

FIG. 13.

This experiment is illustrated in all our books on natural philosophy.

The foregoing diagram, copied from Olmstead's Philosophy, does not show as accurately as Faraday's projection of the lines upon a globe-magnet the comparative distance from the poles of the needle, at which the longitudinal currents commence and terminate, and *where the filings will not adhere* to any considerable extent. The lines shown upon the needle should bear the same proportion to its length as the trade-winds bear to that of the earth, measured from pole to pole, and if the needle had a globular form they would so appear.

These lines are made by currents arising from one side of the magnetic equator, and passing over to the other. Doubtless, just such currents rise, and pass over upon the earth.

Magnetic and electric currents carry the air with them. This is well settled by experiment. *Oxygen*, too, is *magnetic*, and capable both of receiving and retaining polarity and of combining with, or attracting and retaining vapor, and of course the moisture of evaporation. Here then we have a power existing, capable of producing the result—precisely, and with evident wisdom adapted to its production—ever present and active; and no other known agent can.

Is it not then the agent?

Let us look a little further. This result is affected by the action of the sun: the trades with the central belts of rains travel north and south after it; so does the sun affect the magnetic currents every where, even the magnetic needle is daily affected by its action, as it increases the intensity of the terrestrial magnetic currents, and hence its well established diurnal oscillations.

Again, along the eastern lines of the continents which skirt the great oceans on the west, run the northerly and southerly lines of no variation, and of greatest magnetic intensity. Here are the trade currents gathered into a volume, which curve and carry unusual fertility to South-eastern Asia, and North America, and in those great aerial gulf streams we find the *intense* electric action which produces the typhoons of the former, and the hurricanes of the latter. It may still be said that these conditions and phenomena of the trade-wind region, are not produced by magnetism or magneto-electricity, *but the objector can point to no other adequate power*. That it must be heat, electricity, or magnetism, must be admitted. There is no other power known. Heat demonstrably can not produce them. Magnetism or electricity therefore must, and they are doubtless states or phases of the same power, producing in their different states or phases the different results. And even heat—atmospheric temperature, is often, if not always the result of their action. In the present state of science, it is enough for me that the *magnetic longitudinal currents are there*; that they are *lines of force* and *adequate*; that *oxygen is magnetic*, and therefore the atmosphere must be affected by them—that so far as we can reason from analogy, they ought to produce the effect upon the atmosphere which we find produced, and until further light is thrown upon the subject I shall presume that they do. Every step we take hereafter in this investigation will confirm the presumption.

There is one peculiarity to be more particularly noticed before we leave the trade-wind region, and we are now prepared to notice it.

The belt of rains, formed by the currents of the two trades, threading their way through each other—how are they produced? Why should the place where the currents thus pass through each other be a place of almost daily precipitation? There is, in fact, no ascension, except that which the currents have in their line of ascent to attain the elevation which the magnetic law of the current requires.

The trades have passed over an evaporating surface and are charged with moisture. This moisture they hold in magneto-electric combination. *Evaporation* does not depend upon *temperature*. Ice and snow evaporate at all temperatures (Howard, vol. 1, p. 86). So the cold N. W. wind, full of positive electricity, will lap up, as it were, the pools from the earth, with astonishing quickness; and when this electricity is deranging the action of the machinery and material of the manufacturer, he allays it by a supply of moisture, with which the electricity can combine. Nor does the air lose its moisture when below the freezing point. In all parts of the atmosphere, as at the surface of the earth in winter, moisture is held in large quantities in the coldest and severest weather; and it is not till it moderates, and a perceptible *electric* change takes place, that it is precipitated as rain or snow. Doubtless there is an exposure of considerable surfaces, of opposite currents, charged with opposite polarity, and a constant depolarization where their surfaces meet. May there not be a consequent dissolution of the electro-magnetic combination between the air and moisture, or the excitation of that electric action which attends or produces like rains every where? and hence the constant precipitation. This is rendered probable, by the fact that precipitation, at the meeting of the trades, takes place in level countries in the day-time, between 10 A. M. and sunset, in showers, with thunder and lightning, as with us in summer, although among the mountains the rain sometimes falls in the night also. The precipitation in the heat of the day is obviously induced by the action of the sun, although it is by no means certain that the friction of the opposing surfaces does not assist in the operation.

I am well aware that the lines of magnetic force curve upward and carry the trades with them, and that, therefore, precipitation by condensation from the mere cold of the upper stratum of the atmosphere is possible. But, there are three reasons why I do not believe such to be the fact.

1st. Precipitation takes place in the day time mainly, and in sudden, isolated, heavy showers and not in steady continuous rain. Nor is there condensation or continual mist at other hours of the day.

2d. They occur at a time of day when the sun is affecting the magnetic currents most powerfully, *viz.*, between ten o'clock A. M. and sunset, and mainly at the time of greatest heat.

3d. The counter-trades *do not precipitate* after they leave the rainy belt, although at a great elevation, until they reach the outward limits of the trades; and they *do precipitate again*, although they gradually descend *nearer the earth*, as soon as they become subject to the action of the currents of an opposite magnetism. Their precipitation is partial too, even then, and they carry a portion of their moisture through an atmosphere of the coldest temperature up to the geographical poles.

A similar result attends the action of the sun in the extra-tropical regions. Cumuli commence forming in the counter-trade, or at the line between that and the surface current, at the same time of day that the diurnal motion of the magnetic needle commences, or the rain clouds form in the tropics; they continue to enlarge here as there, till about the same hour of the day that the *needle* obtains its maximum diurnal variations; and when the influence of the sun upon the needle ceases, and it returns to its original status, the cumuli disappear. Hail storms too, it is said, always, or generally occur in the day time.

In like manner the sea-breezes and other fair-weather surface winds, rise in the forenoon with the influence of the sun upon the magnetic currents and the needle, and die away at nightfall when the influence ceases.

There are other electro-magnetic, or to speak more correctly, magneto-electric, effects of the sun's action equally illustrative, which tend to show that the precipitation at the passing of the trades, is the result of their action upon each other, aided by the sun, to which we shall allude when we come to speak of the causes and character of the surface winds of the extra-tropical regions.

As, however, this takes place only, or mainly, where the threading surfaces meet, it is but partial, and the body of the respective polarized currents pursue their way unaffected, toward the opposite magnetic pole—and there for the present we leave them.

Storms sometimes originate in these currents, when concentrated, as in the West Indies, the China Sea, the Bay of Bengal, and Indian Ocean, while passing through the rainy belt, and move with the current to the north-west if issuing on the north side of it, and to the south-west if issuing on the south side of it, until they respectively get beyond the extreme limits of the trades, and then they curve to the eastward, imbedded in and following their current. The peculiar extension of the land to the east on the northern portions of South America, prevents the gathering of an aerial gulf similar to the one which we have described to the north-west, entering upon our division of the continent over the Gulf of Mexico. It is otherwise in the Indian Ocean, and there the storms are found issuing from the rainy belt on the southern side, sweeping over the Mauritius and other islands of that ocean, and *often simultaneously* with storms issuing on the north over the Bay of Bengal. Colonel Reid mentions instances and gives a diagram.[2]

These storms in milder forms issue from the rain belt at other points, and may issue any where, but will always be found most extensive and most violent, that is to say, as hurricanes and typhoons, in the concentrated volumes of counter-trade on the western side of the great oceans, within a few hundred miles of the lines of magnetic intensity and no variation, and when they form in the rainy belt they are highly electric. Most frequently, however, as we shall see, they form in these currents after they have issued from the rainy belt, and after they have passed the extreme limits of the trades and become subject to the circular and perpendicular magnetic currents which exist north and south of the longitudinal ones, and which when seen upon the magnetic needle, attract the filings and cause them to

---

[2] Law of Storms, p. 42.

adhere—although but slight attraction or adhesion takes place where the longitudinal currents exist.

Such, then, are the atmospheric arrangements and phenomena of the trade-wind region, and the cause that produces them; such is the character and cause of the enlarged volume of counter-trade, which spreads out and blows over our country as permanently as the S. E. trades blow on the South Atlantic and South America, returning to us the rivers which had run from us to the sea.

CHAPTER VI.

Coming back now, to a consideration of the course and functions of the counter-trade after it leaves the northern limit of the surface-trades, we find it curves to the eastward and gradually assumes about an E. N. E. course, and becomes a W. S. W. current where it crosses the line of no variation, and continues on until it passes off over the Atlantic; and this course and curve is analogous to what may be found true of the counter-trades every where. It is best illustrated by the course of all the storms (in the American sense of the word, as distinguished from thunder showers and other brief rains), which have been traced north or south of the limits of the trades. It was found by Mr. Redfield in most of the storms investigated by him, which originated within, or north of the tropics.

Doubtless it was the actual course of the others, and that the investigation was imperfect. All the great autumnal, winter, or spring storms which have traversed the whole or any considerable portion of the territory of the United States, east of New Mexico, which have been investigated by Professors Espy, Loomis, Redfield, or others, have been found to follow this course. A storm which passed over Madeira, appears from the investigations of Colonel Reid to have followed the same law of curvature.

And so, doubtless, did another which he has described as passing over the Levant. The storms which supply the winter rains of California and Utah, reach them by this law of curvature and progress, after the northern limits of the trades have descended to the south with the sun, so that the counter-trades of the Pacific may descend to the surface and curve in upon them. But the absence of a concentration of the counter-trade, and its deficient action because of its passage over mountain ranges, and their location so near the northern limit of the trades that their storms can not expand and become extensive, as well as their weaker magnetic intensity, prevent their storms from becoming violent, and their supply of rain is not large and much of it falls in showers. The same is true of the Barbary States, of Syria, and Persia, and of Southern Europe; and indeed of all the countries of the globe which lie between the winter and summer extreme limits of the surface-trades, and without the limits of the two concentrated counter-trades. Enough appears in the writings of the meteorologists of Europe to show, that their long continued rains, which are analogous to our storms and are *preceded by the formation of the true cirrus of the counter-trade*, follow the same great law of curvature and progress; although the presence of the Gulf Stream with its mass of south polar waters on the western side of the British Islands, Denmark, and Norway supplies them with showers, and fogs, and cumuli from the west and north-west, and makes the mean of the surface winds of their storms somewhat variant from ours. A like law reversed prevails in the southern hemisphere. The storms of New Holland and the Indian Ocean, south of the limits of the trade, curve to the eastward and travel about south-east, their *south-west* being a *clearing off wind* as our *north-west* is, and *precisely similar in all its other characteristics*, where the relation of magnetic intensity is the same.

The storms of the Pacific on the S. W. coast of South America, in like manner travel to the S. E., flooding the western slopes of the mountain ranges with rain, and aggravated by the intensity of the magnetic currents at the extremity of the continent in a high latitude, meet the mariner in the face as he emerges from under the lee of the land and attempts to pass the Horn. It will ultimately be shown that the precipitation which takes place, as the storms and counter-trades pass north and east in the northern hemisphere and south and east in the southern hemisphere, is owing less to cold than increased magnetic intensity. And all this is the result of one great uniform law, existing every where, varying in its phenomena only in consequence of the difference in volume, and magneto-electric intensity of the portions of the counter-trade, as of the surface-trade at different places, and the different magnetic intensity of the local perpendicular and circular currents of the earth over which they pass, at different periods and at different points.

Mr. Redfield and Lieutenant Maury have assumed that our S. W. current comes from the Pacific Ocean. Aside from the adverse evidence which the investigations of the former in relation to the course of the West Indian storms, and their curving over the continent, furnish to the contrary, and that which has herein before been stated in relation to the law of curvature, it is obvious they are mistaken, for another and conclusive reason.

In order to reach us from the Pacific in a direction from S. W. to N. E., it must pass the table lands and mountain ranges of Mexico and New Mexico, and it would supply them bountifully, even if it did not thereby leave us comparatively rainless and sterile. Every where currents passing from the ocean *over mountain ranges* part with a large share of their moisture. Thus the counter-trade which curves over the Andes and over Peru, is deprived of its moisture and leaves the western coast rainless. So in degree of the counter-trade which curves over the Himalaya and Kuenlon Mountains, and from there passes over the Desert of Cobi, to the north and east—it is deprived by those elevated ranges of its moisture. So the mountains on the south-western coast of South America are drenched with rain, while Patagonia, which lies on the east of them is comparatively dry. And so of every other country similarly situated.

Now the mountain ranges and table lands of Mexico are not thus supplied with moisture. For the space of four months in Southern and less in Northern Mexico, and in summer, and while the belt of the tropics is extended up over them, they have rain and in daily showers which *travel up from the south*, indicating the course of the counter-trade. (See Bartlett's Personal Narrative, vol. ii. p. 286.) At other seasons, and while we are bountifully supplied, they are dry. In short, there are no two portions of the earth that differ more widely in regard to their supply of moisture, and all their climatic characteristics and relations. It is therefore, according to all analogy, impossible that our counter-trade should come from the South Pacific across the continent and below 35°, and in this also those gentlemen are mistaken.

Messrs. Espy and Redfield recognizing the existence of "a prevailing" S. W. current, but considering the surface-winds beneath it as the principal actors in producing the atmospherical conditions and changes, have attributed no office to that current, except that of giving direction and progression to our storms. This is their great mistake. It plays no such unimportant part in the philosophy of the weather, as we have already incidentally seen, and will proceed still further to consider.

All our storms originate in it. This we may know from analogy.

Where there is no counter-trade, outside of the equatorial belt of rains, and within influential distance of the earth, there are neither storms nor rain. So, when, as we have seen, the concentration of the volume of northern counter-trade in the West Indies, gathered by the hauling of the S. E. trades more from the east, as they approach the central belt, diminishing the volume of the counter-trade over the North Atlantic, the calms and drought of the horse-latitudes are found. And when the counter-trade is small in volume and weak in intensity, by reason of the fact that the surface-trades from the opposite hemisphere which constitute it, formed upon land where evaporation was small, as upon Southern Africa and New Holland, or formed where the magnetic intensity was weak, or passed over mountain ranges in their course, the annual supply of rain, the ranges of the barometer, and the alternations of atmospheres conditions are remarkably less.

We have already seen where the rainless portions of the earth are, and why they are so; because those lying north of the northern limit of the equatorial rainy belt were yet too far south to be covered by the line of extra-tropical rains; or in other words, too far south to be uncovered by the surface N. E. trades and the longitudinal magnetic currents, and to be covered by the counter-trades in contact, or nearly so with the earth, and influenced by the perpendicular north polar magnetic currents. Thus we have seen that the rains of Southern Mexico were summer rains, due to the northern extension of the equatorial rainy belt; those of California were winter rains, due to the southern extension of the extra-tropical rains following the N. E. surface trades. We have also briefly alluded to the fact that either side of the equatorial rainy belt, evaporation is going on for months under a vertical sun, without precipitation—unless it be from an occasional brief storm of great intensity which originates in that belt at the line of it, and passing on in the counter-trade, reverses, for the time being, by its concentrated and powerful action, like a magnetic body introduced into the field of another magnet, the surface-trades. Mere evaporation then, does not produce the storm, or shower, or rain, where most active in the dry torrid zone. It may be said that those dry portions are, for the time being (as the rainless portions of the earth are continually), within the operation of the surface-trades, and that therefore the evaporated moisture is carried away by them toward the equatorial rainy belt. Precisely so; but why carried away? Why should it not condense, occasionally, at least, and drop the rain as it passes along, if a great supply of moisture from excessive evaporation could furnish rain. Perhaps it may still be said it is going from a cold to a warm section. This is not true, as we have shown.

But, it may be said that the rainless regions at any rate receive no moisture, and therefore can not supply any by evaporation. This would not meet the case, as it would still be true that when the rainy belt has left a given spot, the dry weather sets in with excessive evaporation, and the north-east trades in summer, blowing from the countries lying north of the rainless regions, and which have been supplied during the interval by the extra-tropical rains, and are loaded with evaporation, are passing over the rainless regions on their way to enter the central belt. So blow the N. E. trades from the Mediterranean, and the Barbary States *over the Desert of Sahara* and into the rainy belt south of it; but drop no moisture on their way, because exposed to no magnetic currents of an opposite polarity.

But it is not true that all the rainless regions are without evaporation. Egypt is an exception. The annual freshets of the Nile saturate its central valley, and vast reservoirs of water are saved from it and let out over its surface, and it all evaporates, but produces no rain. And so are large quantities turned aside and scattered over the bottom lands of Northern Mexico, and other countries, during the dry season, and their evaporation furnishes no rain. Hygrometers and dew points are of no consequence there—nor are they of any, on either side of the rainy belt, where six perpendicular feet of moisture is evaporated in six months.

Again we have alluded to a strip of coast on the Pacific west of the mountain ranges of South America, lying partly in Peru, partly in Bolivia, and partly in Northern Chili, which, although long and narrow, washed by the broad Pacific Ocean, is without rain. South America has no other *wholly* rainless region, so far as is known. A part of this region would lie between the equatorial belt of rain, and the southern extra-tropical one, and never be covered by either; but the volume of N. E. trades from the Atlantic, although from the make of the land not concentrated to so great an extent as the volume of S. E. trade on the north, and therefore not so liable to hurricanes and other violent storms, is yet sufficiently so to carry the southern line of the equinoctial rainy belt down in winter to the summer line of extra-tropical rains, and give a supply of rain to all the continent—leaving no strictly rainless region south of the equatorial rainy belt and east of the Andes. Those mountains, however, present a barrier to its south-western progress which it doubtless passes to some extent, but deprived of its moisture, and unable to supply the rainless coast region of Peru, Bolivia, and Northern Chili. There is, therefore, a portion of this rainless line of coast which is within the region of extra-tropical rains, over which a portion of the N. E. trades of the Atlantic, as a counter-trade, should or do, curve, and where there should therefore be extra-tropical rains. It is washed by the Pacific, an evaporating surface, and westerly and south-west breezes are drawn in from that ocean over it. Why then is it rainless? The only reason which can be assigned why rain does not fall there is that the high mountain ranges of the Andes intercept and perhaps in part divert the counter-trade, and deprive that portion of it which passes them, of its moisture, by that reciprocal action of opposite polarities which takes place whenever and wherever the trade approaches so near the earth; and it curves over the narrow line of coast with the feeble condensation, and imperfect forms, and varied coloring which mark so peculiarly the rainless clouds of that region. (See Stewart's Journal of a Voyage to the Sandwich Islands, page 72.)

Again, it is estimated, and on reliable data, that twelve perpendicular feet of water are annually evaporated from the surface of the Red Sea, between Nubia on one side, and Arabia on the other; yet they are both rainless countries, except so far as the inter-tropical belt of rains extends up on to a small portion of them. The moisture of evaporation, floated up from a surface covered by the surface-trade is invariably so combined as to remain uncondensed till it has passed south into the equatorial rainy belt, and over to the opposite hemisphere, and been exposed to the currents of an opposite magnetism.

Again, the N. E. trades extended up in summer over the Mediterranean Sea, an evaporating surface, blow over the Barbary States in June and July, but furnish no rain. And so of the S. E. or N. E. trades which blow over Brazil and other countries in the absence north or south of the tropical belt of rains.

It is obvious from these facts—and more like them might be cited—that mere evaporation, however copious or long continued, does not make the storm or shower in the locality where it takes place, and *without the existence and influential agency* of a counter-trade; and that *reciprocal action*, whatever it may be, that takes place *between it and the earth*.

Again, our own experience is conclusive of this. We have no surface-trade north of 30°, and yet a long drought and great evaporation may follow a wet spring. Belts of droughts and frequent rains occur every year in different portions of the country side by side, and *the dividing line follows the course of the counter-trade*, and is sometimes distinctly marked for weeks. When a change occurs in the counter-trade, whether from causes existing there or the influence of terrestrial magnetism (in relation to which we shall inquire hereafter), showers form or storms come on: until it does they will not. Efforts at condensation will occasionally appear, but they will be feeble and ineffectual, and occasion a repetition of the axiom that "all signs fail in a drought." And we may know it from direct observation.

The first indications of a storm, and of most if not all showers, are observable in the counter-trade. These indications, so far as they are visible, are of course to be looked for in the west; although the direction and character of the surface-winds are often indicative of these changes when not visible at the west as we shall see.

The indications are those of condensation, and vary very much in different seasons of the year. It is not my purpose in this place to examine them particularly. They will be alluded to hereafter under the head of prognostics. Suffice it now to say, then, that whether it be the long threads or lines of cirrus which occur in the trade in the winter after a period of severe cold, following the interposition of a large volume of N. W. cold air and the elevation of the counter-trade; or the forms of cirrus which occur at other times and other seasons; or whether it be the ordinary bank at night-fall, or the evening condensation which makes the "circle" around the moon, or the morning cirro-stratus haze which gradually thickens, passes over and obscures the sun, all which may be followed by the easterly scud and winds: they are alike condensation in the trade, the advance or forming condensation of a storm or showers.

The state of the weather, whether hot or cold, is extensively affected by this trade current. As we have already suggested, the mere presence of the sun in its summer solstice, or its absence in winter, is not an adequate cause of all the sudden and various changes to which we are subject. The state of the counter-trade, which is always over, or *within influential distance of us*, and sometimes probably in contact with us—the nature of the surface-winds which it is at any given time creating and attracting around us, and the electric condition of the surface-atmosphere *induced* by it, or by the immediate action of the earth's magnetism, produce those sudden changes which mark our climate. When no intervening surface-winds elevate it above us, and there is no storm or other condensation within influential distance, it induces the gentle balmy S. W. wind of spring—the cooling S. W. wind of summer—the peculiar Indian summer air of autumn, or the comparatively moderate, although cold, open weather of winter. If there be a partial tendency to condensation in it, the cumuli form under the magnetic influence excited by the sunbeams from ten to three o'clock in the day, and float gently away to the eastward, disappearing before night-fall. If the disposition to condensation is stronger, whether inherent or induced by an increased local activity of terrestrial magnetism, these cumuli will increase toward night-fall, or earlier, and terminate me showers; and if it is in a highly electrical state, the still oppressive sultriness which precedes the tornado, and that devastating scourge may appear. If this disposition to condensation becomes extensive, cirri form and run into cirro-stratus, or they extend, coalesce, and form stratus; the surface-wind will be attracted under them, the thermometer fall in summer or rise in winter, and a storm begin. Intense action and sudden cold may exist in and under this counter-trade over the southern portion of the country, while all is calm, warm, and balmy at the north. Heavy snow storms sometimes pass at the south when there are none at the north, and a corresponding state of the weather follows. If a large body of snow fall at the north, the winter is cold, regular, and "old fashioned;" if little snow falls at the north and more at the south, the winter at the north is open and broken. I have known the ice make several inches thick at Baltimore and Washington, when none could be obtained for the ice-houses on the Connecticut shore of Long Island Sound. In short, although heat and cold are mainly dependent upon the altitude of the sun, aided by the other arrangements we have alluded to, yet the counter-trade, and the reciprocal action which takes place between it and the earth, are most powerful agents, mitigating the rigors of winter, bringing about the changes from cold to warm weather which the sun is too far south to produce. And on the other hand, by this reciprocal action, producing the electrical phenomena, the gusts, the tornadoes, the hail storms, and the cool seasons of summer, and the period of intense cold in winter.

All our surface-winds, except the light, peculiar W. S. W. wind which is felt where the counter-trade is in contact with the earth, and which is a part of it, and perhaps the genuine N. W. wind which is very peculiar, are incidents of the trade, and are due to its conditions and attractions. We have already said this was true of the easterly wind and scud of a storm—it is alike true of all. The storm winds east of the Alleghanies are usually, though not always, from the eastward. They are sometimes from the southward, as they doubtless are still more frequently in the interior of the continent.

There is occasionally a southerly afternoon wind, followed by short rains in spring and fall, or a succession of showers in summer, which is rather a precedent wind than a storm wind; blowing toward and under an advance portion of the storm at the north, and hauling to the eastward when the rain sets in, or to the westward when the showers reach us.

When there are no storms, or showers, or inducing electric action in the counter-trade, within influential distance to disturb the surface atmosphere, it is calm. If a storm approaches, or forms within inducing distance, the surface atmosphere is *affected* and *attracted toward the storm*, from one or more points, and "blows," as we say, toward and under it. It commences blowing first nearest the storm, and extends as the storm travels, or becomes more intense and extends its inducing influence. I have repeatedly noticed this in traveling on steamboats and railroads running *toward* or *from*, and in several instances *through* a storm, and telegraphic notices and other investigations prove it. The point from which the surface atmosphere is attracted and blows, depends very much upon the position of the storm in relation to bodies of water and the point of observation, and its shape; and the force with which it may blow will depend much upon its intensity.

Let us take an instance or two by way of illustration of all these points; and as I have given instances of summer in the introduction, we will take those of winter. It is January of an "old fashioned winter;" the snow is about three feet deep in Canada, about one foot in Southern New York, and a few inches in Philadelphia, and so extends west to the Alleghanies at least. For several days the sky has been clear, the thermometer rising in the day-time, in the vicinity of New York to about 25° Fahrenheit, falling at night to about 6°, with light airs from the N. W. during the middle and latter part of the day; the counter-trade and the barometer both running high; cold but pleasant, steady, winter weather. There is a warm south-east rain and thaw coming, as one or more such almost invariably occur in January. How coming? The sun is far south, and shines aslant, but through a pure and windless atmosphere; he has tried for several days to melt the snow from the roof; a few icicles are pendant from the eaves; but the body of the snow is still there. How can a thaw come? not from the sun, surely. No, indeed, not from the action of the sun directly, upon our country, nor from the Atlantic or the Gulf Stream which is off our coast. But a portion of the current of counter-trade is coming, heated by his rays and the warm water in the South Atlantic, in an intense magneto-electric state, capable of inducing an electro-thermal change in the surface atmosphere which it approaches, and of being reciprocally acted upon by the north polar terrestrial magnetism. It is now over Northern Texas and Western Louisiana, it will be here day after tomorrow. The day passes as the day previous had passed; the sleigh-bells jingle merrily in the evening; the moon shines clear all night; the storm is coming steadily on, but its influence has not reached us, and the morning and midday are like those which preceded it. As nightfall approaches, however, the thermometer does not fall as rapidly as on the day previous; the sun shines dimly and through lines of whitish cirrus cloud extending from the horizon at the west, appearing darker as the sun descends and shines more *horizontally* through them — perhaps mainly in the N. W. — and which extend up and over toward the E. N. E. The air next the earth begins to feel raw; it is changing, not from warm to cold, but *electrically* from positive to negative; and dampening, from a tendency to condensation by induction, as we shall see — the same condensation which in warm weather may be seen on flagging stones, and walls, and vessels containing cold water. The advance cirrus condensation of the storm is over us and affecting us; the earth too is affecting the adjacent atmosphere by action extended from beneath the storm. Still there is no wind, although sounds seem to be heard a little more distinctly from the east, and so ends the day. Evening comes, and the moon wades in a smooth bank of cirro-stratus haze, with a very large circle around her; the cirrus bands of haze have coalesced and formed a thin stratus. The storm is coming steadily on, its condensation is seen to be thicker as it approaches, it is now raining from one hundred to one hundred and fifty miles to the west, but we do not know it.

That it is about to storm all believe, for all are conscious of a change. The candle if extinguished will not relight as readily, if at all, on being blown; there is a crackling almost too faint for snow in the fire; the sun did not set clear; the old rheumatic joints complain, and the venerable corns ache.

Morning comes, and the storm is on. The wind is blowing from the S. E., the scud are running rapidly from the same quarter to the N. W., the thermometer continues rising, and it rains. The storm has reached us and the thaw has commenced. Gradually, as the densest portion of the storm cloud reaches us, it darkens; the scud are nearer the earth, and run with more rapidity; the rain falls more heavily and continuously, and by the middle of the day a thick fog has enveloped the earth; the wind is dying away, and the trade itself, with its southern tendency to fog, has settled near us; the barometer has fallen, the thermometer is up to fifty degrees, the water is running down the hills, the snow is saturated with water and is disappearing under the influence of the fog, the rain, and the warm air. Evening comes; the south-east wind and the rain have ceased; the rain clouds have passed off to the eastward; the fog has followed on and disappeared; there is a light trade air from the S. W.; the moon shines out, and a few patches of stratus, broken up into fragments and melting away, are following on in the trade: the storm is past.

Hark! to the tones of Boreas as he bursts forth from the N. W., and rushing, whistling, howling, dashes on between the trade and the earth, following the storm. Now the barometer rises rapidly, the thermometer falls, and in an incredibly short time all is congealed, and cold and wintery as before. The cold N. W. wind has again interposed between the trade and the earth; the trade is elevated a mile or more above it and is entirely free from its influence and from condensation; the deep blue of a sky "as pure as the spirit that made it" is over us, and steady winter reigns again.

It is obvious that there was nothing in the action of the sun upon our snow-clad country, to induce the thaw or the storm. It began, continued, approached, and passed off to the N. E. in the counter-trade. The S. E. wind which existed every where within its influence: in the interior States, Missouri, Illinois, Indiana, Ohio, Michigan, and in Canada, as well as upon the Atlantic coast, commencing in the former earlier than upon the last, was the result of its induction and attraction. Of the N. W. wind that followed we shall speak hereafter. If any one doubts whether this be a true sketch let him examine the investigation of a storm published by Professor Loomis, or observe for himself hereafter. If, however, the storm of Professor Loomis is referred to, it should be remembered that his notes show the occurrence of a slight distinct snow storm at the N. W. stations one day in advance of the principal storm. The latter appears first as rain at Fort Towson, on the nineteenth, moving north and curving to the east—its center passing near St. Louis, and south of Quebec, and the whole storm enlarging as it advanced.

Take another instance. Since the thaw it has not been quite as cold as before; but the rain-soaked snow is hard and solid, the ground, where the snow was blown or worn off, icy and slippery—the thermometer falls during the night to about 12°, and rises to about 30°; the sun makes no impression upon the snow; the firmament is of the deepest blue, the borealis at night vivid. "O, for a storm of some kind, to mitigate the still severe cold;" for the thaw has made us more sensitive, and storm winds do blow warm in their season. But patience, it will come. Another day, or two, perhaps, pass: the sun rises as usual, the thermometer has the same range still. "Long cold snap," we exclaim; "how long will it last?"

A change is coming, but this time it will snow. About an hour or two after sunrise the cirrus threads are discoverable again in the west, but now they are most numerous in the S. W. As the day passes on they thicken and advance toward the E. N. E., the sun begins to be obscured, the thermometer rises, and it slowly "*moderates*." There is a snow storm approaching from the S. W.

But the thermometer rises slowly; it must get up to 26° or 28° before it can snow much. I have known in one instance, at Norwalk, a considerable fall of snow, although much mingled with hail, when the thermometer stood at 13° above zero, and one, a moderate fall, some two inches, with it at 24°, but these were exceptions. The snow range of the thermometer on the parallel of 41° north latitude, and south of it, is from 26° to 30° above 0°; when colder or warmer it may snow to whiten the ground, or perhaps barely cover it, but usually rains or hails. We have seen that in the polar regions, according to Dr. Kane, it is about zero, but the rise of the thermometer there, previous to the snow, was about the same as here, *i. e.*, from 15° to 25°. This fact is instructive. Since the foregoing was written, and on the 7th of February, 1855, a snow-storm of considerable length set in, with the thermometer at 5°, and continued more than twenty-four hours, the thermometer gradually rising. The snow was very fine, like that described by Arctic voyagers as falling in extreme cold weather.

As the dense and darker portions of the storm approach, and although the sun is obscured, and the ground frozen, it continues to moderate, and at evening, when the thermometer is up to 28°, and the dense portion of the storm has reached us, gently and in calmness the snow begins to fall. Perhaps a light air following the storm, or the presence of the trade near the earth, at first inclines the snow-flakes to the eastward. This is frequently so at the commencement of snow storms. Ere long, however, the wind rises from the N. E., and the snow is driven against the windows, rounded and hardened by the attrition of its flakes upon each other, in their descent through the eddying and opposite currents. The next day we rise to witness a heavy fall of snow, perhaps, and a continued driving N. E. storm, in full blast; the snow whirling and settling in drifts under the lee of every fence or building.

Can it be, you ask, that this driving wind is but an *incident* of the storm? the result of *attraction*, while the storm clouds are sailing quietly and undisturbed on in the counter-trade above, directly over the gale which is blowing below? It is even so. Nor has it "backed up," as it is termed by those who have ascertained that it has commenced snowing first, and cleared off first, at a point west of them. You saw, or might have seen, the cirro-stratus cloud passing to the E. N. E. in the afternoon, and until the snow-flakes filled the air, and the clouds became invisible. You may still see that the wind will die away before the storm breaks, and "come out" gently from the S. W., unless it should back into the northward and westward, and in either event you may see the last of the storm clouds, as you did see, or might have seen the first of them, pass to the eastward. Toward night the wind dies away, and the storm passes off abruptly, or the sky becomes clear in the N. W. Now you may see the smooth stratus storm cloud, continuous, or breaking up into fragments and passing off to the east, even at the edge which borders the clear sky in the west or north-west, to be followed that evening or the next day, by the north-west wind and its peculiar fair-weather scud.

I have given these as instances illustrating the manner in which rain and snow storms originate the surface easterly winds in winter.

But it must not be supposed that they commence with precisely the same appearances in every case in winter; much less in summer. There is very great diversity in this respect, in different seasons, and in different storms during the same season. A great many different and accurate descriptions might be given, if time and space would permit, which all would recognize as truthful. Very frequently in summer, and sometimes in winter, the wind will set in from the eastward, and blow fresh toward a storm, before the condensation in the trade, which forms the eastern and approaching edge of the storm, has assumed the form of a distinct cloud. Not unfrequently, when it is calm next the surface, a narrow stratum of easterly wind, a half a mile or a mile above the earth, may be seen with a continuous fog, condensing, but not in considerable patches like the usual scud, running with great rapidity toward the storm. Such a stream of fog blew with great rapidity for thirty-six hours toward the storm which inundated Virginia and Pennsylvania, in 1852, and carried away the Potomac bridge at Washington. Such a stream of fog was visible the evening before the great flood of 1854, which inundated Connecticut, and curried away so many railroad and other bridges. I have also seen such a stream of fog running at about the same height, when it was calm at the surface, from the S. W. toward a violent storm which formed over central New England—and from the north toward a heavy storm passing south of us. Such strata form, as far as I have been able to discover, the *middle current* of storms which are accompanied with very heavy falls of rain. These double currents are much more common than is supposed. East of the Alleghanies, short and heavy rain storms, which commence north-east, hauling to the south and lighting up about mid-day *after a very rainy forenoon*, frequently have a S. E. or S. S. E. middle current of this character, which involves the whole surface atmosphere when the storm has nearly passed, and the N. E. wind dies away, and the wind seems to haul to the S. S. E. and S.; so that it is rather the prevalence of a *different* and *coexisting current*, than a hauling of the *same wind*, which marks the period of lighting up in the south.

Sometimes the easterly wind will set in and blow a day or two before the border of the storm reaches us. Sometimes the storm is passing, or will pass, in its lateral southern extension, south of us, and the condensation in the trade extends over us sufficiently dense to induce an easterly current beneath it, but not dense enough to drop rain, and then we have a dry north-easter. I can not, within the limits I have prescribed, allude to all the peculiarities attending the induction and attraction of an easterly wind, by the storm in the counter-trade. They are readily noticeable by the attentive and discriminating observer, and their existence and cause is all with which I have to do at present.

Winds from the north, or any point from N. N. E. to N. N. W., are comparatively infrequent in the United States, east of the Alleghanies—though it is otherwise in the vicinity of the great lakes.

Sometimes the wind "backs," as sailors term it, during a N. E. storm, from the N. E. through the N. N. E., N., and N. N. W. to N. W. When this takes place, it is toward the close of the storm. Occasionally, though very rarely, it continues to storm after the wind has passed the point of N. N. E., and until it gets N. W. I have known a few instances in the course of thirty years, and but a few. They are exceptions—rare exceptions. When the wind thus backs from the N. E. to the N. W. through the N., you may be very certain that the body of the storm, or at least the point of greatest intensity and greatest attraction, is at the time passing to the southward of you. This is most commonly the course of the wind when the storm extends far south and lasts several days, and does not extend north far, or if so, with much intensity, beyond the point of observation. The change of the wind is explained by the situation of the focus of intensity and attraction, to the south of the observer, and its passage by on that side.

Probably in locations further north and (as I think I have observed) south of the lakes, it may be more frequent than upon the parallel of 44° east of the Alleghanies (which is as far north as I have observed), inasmuch as the further north the locality, the more likely storms and other disturbances in the counter-trade will be to pass to the southward of it.

Between the N. E. and S. E. the wind may blow from any point, before and during storms, and in a clear day in the morning, as a light variable breeze, or, after mid-day, toward approaching showers. I have known it blow all day during a storm from due east; to change back and forth between south-east and north-east, and to blow for hours from any intermediate point—as different portions of the storm were of different intensity, and exerted a more or less powerful inducing influence; and doubtless this often takes place at sea. It depends upon the situation of the focus of attraction of the storm, its shape relative to the particular locality, and with reference to the atmosphere east of it, and peculiar local magnetic action; or, as is sometimes the case in low latitudes, is owing to the fact that the storm is made up of many imperfectly connected showers, which have different force, and induce changeable and baffling winds.

The inducing and attracting influence of the approaching storm is exerted sooner, and with most force, upon the surface atmosphere, over bodies of water like the ocean and the lakes. Thus, the wind will set from the eastward toward an approaching storm out upon Long Island Sound, for hours before it is felt upon either shore; and when all is calm in the evening on land, and often before the moon forms a halo or circle in the milky condensation of the approaching storm, or any sign of condensation is visible, the breaking of the waves upon the shores may be heard. Doubtless this may be observed on the shores of the Atlantic at other points.

This power of attracting the surface atmosphere from bodies of water like the ocean and the great lakes, will account for two apparent anomalies, mentioned by Mr. Blodget in a valuable and instructive article read to the Scientific Convention, in 1853, regarding the annual fall of rain over the United States.

First—the influence of mountains in extracting the water from the atmospheric currents which pass over them, is well known and readily explainable. Mr. Blodget, however, found that the source of our rains, whatever it might be, when it reached the Alleghanies, was so far exhausted of its moisture that those mountains extracted less from it than fell to the westward, by some five to ten inches annually; and that the fall of rain upon them was less than upon the Atlantic slope eastward of them, to the ocean. This does not accord with observation elsewhere, but is easily explained. As the storm approaches the ocean, it attracts in under it the surface atmosphere of the ocean, loaded with vapor, condensing in the form of fog and scud, as it becomes subject to the increasing influence of the storm. Although the scud and fog would not of itself make rain, it aids materially in increasing the quantity of that which falls through it. The drops, by attraction and contact, enlarge themselves as they pass through, in the same manner as a drop of water will do in running down a pane of glass which is covered with moisture. The small drop which starts from the upper portion of a fifteen-inch pane, will sometimes more than double its size before it reaches the bottom. *It is by this power of attracting the surface atmosphere, which contains the moisture of evaporation, under it, and inducing condensation in it, that the moisture of evaporation which rarely rises very far in the atmosphere is made to fall again during storms and showers.* This attraction of a moist atmosphere from the ocean accounts for the excess of rain on the east of the Alleghanies, compared with its fall upon them. So the great valley of the Mississippi is comparatively level, and less of its water runs off than of that which falls upon the Alleghanies. There is, therefore, more moisture of evaporation in the atmosphere of the former to be thus precipitated and add to the annual supply of rain upon that valley, and it exceeds that which falls upon the Alleghanies. Those mountains, too, are elevated but about 1,500 feet above the table-lands at their base, and exert little influence on the counter-trade. If they, were 6,000 or 8,000 feet high, a different state of things would exist.

Second—Mr. Blodget found the quantity of rain which fell in Iowa, and to the south and west of the lake region, to be greater than fell over the lake region itself. This is doubtless in part owing to the same cause. The counter-trade, in a stormy state, attracts the surface atmosphere from the lake region, with its evaporated moisture, before it arrives over it, and therefore more rain falls S. W. of the lake region than upon it. This power of attracting the surface wind of the ocean in under it, produces the heavy gales which affect our coast, and which are rarely felt west of the Alleghanies to any considerable degree; and a storm coming from the W. S. W., extending a thousand miles or more from S. S. E. to N. N. W., may have the wind set in violently at S. E. on the *southern coast first*, and at later periods, successively, at points further north, and thus induce the belief that the storm traveled from south to north.

Mr. Redfield finding that some of the gales which he investigated, particularly that of September 3d, 1821, did not extend far inland, and commenced at later periods regularly, at more northern points, concluded that the gale traveled along the line of the coast to the northward. In this, and in relation to the storm of 1821 (and perhaps some others), he has been deceived. My recollections of that storm are accurate and distinct. But I shall recur to this again when I come to speak of his theory.

Toward storms, or belts of showers which would be storms if it were not summer and the tropical tendency to showers active in the trade, which pass mainly to the north of us, or commence north and pass over us, condensing south while progressing east, the wind may commence blowing before the body of the storm reaches us, from any point between south by west and south east, particularly in the summer season and in the afternoon. When the rain in a storm of this character sets in, in the night, it will sometimes haul into the S. E., if the focus of attraction be situated north of us, and so remain until just before the storm is to break.

There are, however, a class of southerly summer winds which deserve more particular notice. For two or three months in the year—say from the middle of June to the 20th of August—storms on the eastern part of the continent, except in wet seasons, are rare, and most of our rain is derived from showers. During these periods belts of drought are frequent, sometimes in one locality, and sometimes in another, extending with considerable regularity from W. S. W. to E. N. E. in the course of the counter-trade, while rain falls in frequent and almost daily showers to the northward or southward of them. If the daily rains are at the north, over the belt of drought, S. S. W. and S. W. by S. winds blow, sometimes with cumuli or scud, during the middle of the day and afternoon, to underlie the showery counter-trade on the north of the line of drought. Thus, sometimes nearly every day for several days, the evaporated moisture of the dry belt will be carried over to increase the store of those who have a sufficient supply without. During the latter part of the afternoon the clouds in the west may look very much like a gathering shower, but the attractions of the counter-trade fifty or one or two hundred miles to the north, will absorb them all, and at nightfall the wind will haul to the S. W. on a line with the counter-trade, and die away.

If there be a drought on any given line of latitude, and frequent showers or heavy rains at the south of it, although there may not be a like surface-wind, with cumuli and fog, blowing from the north toward it, yet a general, gentle set of the atmosphere, from the N. N. W., or N. W., or other northerly point, toward the belt of rains, some distance above the earth, will often be observable, with a barometer continually depressed, and perhaps a cool atmosphere.

During set fair weather, when the attracting belt of rains is far north, on the north shore of Long Island Sound, the wind, like a sea breeze, will set in gently from about S. S. E. or S. by E. in the forenoon, blowing a gentle breeze through the day, and hauling to W. S. W. on a line with the trade at nightfall, and dying away. During a drought I have known this to happen for seventeen successive days. It is obvious to an attentive observer that this is the result of the influence of the sun in exciting the magnetic influence of the earth, and producing a state of the trade not unlike that which induces the formation of cumuli, and which attracts the surface atmosphere from the Sound in over the land: for the *tendency to cumulus condensation precedes the breeze,* and the breeze is often wanting in the hottest days where no such tendency to the formation of cumuli exists. The same is true of sea breezes elsewhere. They do not blow in upon some of the hottest surfaces. Where they do exist, they do not always blow, but are wanting during the hottest days; and careful observers have identified their appearance with the formation of cumuli, or other condensation, upon the hills inland. They are not, therefore, the result of ascending currents of heated air.

The received theory regarding sea and land breezes is a mistaken one in another respect. There is no such thing as a land wind corresponding in force to, and the opposite of, the sea breeze—occasioned by the comparative warmth of the ocean. These breezes blow mainly within the trade-wind region. Of course they are either beneath the belt of rains or the adjoining trades. They are said to be, and doubtless are, most active and strongly marked on lines of coast, particularly the Malabar coast, and where the trade-winds are drawing usually from them. In the day-time, when the action of the sun increases the action of the magnetic currents upon the land, or there are *elevations inland* which approach the counter-trade, and especially if it is elevated near the coast, as the Malabar coast is by the Ghauts, the attraction of this atmosphere over it *reverses the trade*, or inclines it in upon the land, and it blows in obliquely or perpendicularly, according to the relative trending of the coast and the direction of the surface-trade. Thus, where islands are situated within the range of the trades, the latter will be *reversed* during the day on the *leeward* side, but continue to blow as land winds during the night. So they are sometimes deflected in upon the land on the sides, during the day, and in like manner return to their course in the night. So, too, the north-east trades of Northern Africa, are occasionally (though feebly where the coast is flat) deflected during the day-time, and blow in as N. W. winds. Upon the southern coast of Africa the S. E. trade is deflected, and blows in as a S. W. wind. Upon the south-western coast of North America, the N. E. trades are deflected in like manner, and so are the S. E. trades upon the western coast of South America. Where the coast mountain ranges are very elevated, as upon the western coast of the American continent, this attracting influence and consequent deflection extends to a considerable distance seaward, and hence the westerly winds of California, etc. It must be understood that we are now speaking of the winds which blow within the range and during the existence of the trade-winds or the presence of the dry belt—for the trades are not always perceptible on the land. Captain Fitzroy thus describes the sea breezes of the western coast of Peru, at 23° south latitude. "The tops of the hills on the coast of Peru are frequently covered with heavy clouds. The prevailing winds are from S. S. E. to S. W., seldom stronger than a fresh breeze, and often very slight. *Sometimes during the summer, for three or four successive days, there is not a breath of wind, the sky is beautifully clear, with a nearly vertical sun.* On the days that a sea breeze sets in, it generally commences about ten in the morning, then light and variable, but gradually increasing till one or two in the afternoon. From that time a steady breeze prevails till near sunset, when it begins to die away, and soon after the sun is down there is a calm. About eight or nine in the evening *light winds* come off the land, and continue till sun-rise, when it again becomes calm until the sea breeze sets in as before."

To illustrate this further, I take the following letter from Professor Espy's Philosophy of Storms:

Clinton Hotel, N. Y., Dec. 20, 1839.

To Professor Espy,

Dear Sir,—Understanding you are desirous of collecting curious meteorological facts, I take the liberty of communicating to you what I saw in the month of December, 1815, at the Island of Owhyhee. I lay at that island in the Cavrico Bay,[3] in which Captain Cook was killed, three weeks, and every day during that time, very soon after the sea breeze set in, say about nine o'clock, a cloud began to form round the lofty conical mountain in that

---

[3] Kearakakua Bay (called Cavrico above), is on the S. W. side of the island, and the trade was reversed during the day by the cloud condensation inland.

island, in the form of a ring, as the wooden horizon surrounds the terrestrial artificial globe, and it soon began to rain in torrents, and continued through the day. In the evening the sea breeze died away and the rain ceased, and the cloud soon disappeared, and it remained entirely clear till after the sea breeze set in next morning. The land breeze prevailed during the night, and was so cool as to render fires pleasant to the natives, which I observed they constantly kindled in the evening. I was particularly struck with the phenomena of the cloud surrounding the mountain, when none was ever seen in any other part of the sky, and none then till after the sea breeze set in, in the morning, which it did with wonderful regularity. The mountain stood in bold relief, and its top could always be seen from where the ship lay, above the cloud, even when it was the densest and blackest, with the lightning flashing and the thunder rolling, as it did every day. I passed up through the cloud once, and I know, therefore, how violently it rains, especially at the lower side of the cloud. This rain never extends beyond the base of the mountain;[4] and all round the horizon there is eternally a cloudless sky. The dews, however, are very heavy, and there seems to be no suffering for want of rain. That this state of things continues all the year, I have no doubt, from what an American, by name Sears, who had spent four years there, told me; he had seen no change in regard to the rain.

CALEB WILLIAMS.

Providence, R. I.

Similar citations might be made to show that the sea breeze is induced by the same cause which forms the clouds over the land—that it is frequently wanting for three or four days under a vertical sun, and that the land breeze blows gently and not with corresponding force where there is no surface trade, or where it is deflected, not reversed.

A succession of showers passing across the country to the north, within one hundred to one hundred and fifty miles, almost always produces a southerly wind to the southward of them. There is more that is peculiar about these belts of showers. Although they consist of large highly-electrified cumuli, there is a strong tendency to cirro-stratus condensation in the lower part of the trade over them; and it is that condensation rather than the cumuli, which attracts the surface atmosphere from the south. They would be storms, if the atmosphere had not a summer-tropical tendency to showers. There is, too, a tendency in these belts to extend to the south, and it is generally, as far as I have observed, the extension southerly of those belts, by the formation of new showers which terminate the "hot spells" or "heated terms" of mid-summer. The very oppressive and fatal one of the summer of 1853, was, in character, a type of all—although exceeding them in severity. The first three or four days were calm, hot, and smoky—an appearance which attends

---

[4] Lieutenant Wilkes spent twenty days upon the top of this or an adjoining mountain, and his observations there will be alluded to in another connection.

all similar periods more or less, refracting the red ray of the light, and giving the sun a peculiar dry-weather, red appearance. (This smoky haze is usually atmospheric, and occasionally seen even in March, although not unfrequently fires in the woods fill the air with actual smoke, and very much increase it, and when this is so, the odor of the smoke is often perceptible.) Then we began to have a fresh south-west by south breeze in the day-time, hauling to the south-west, and dying away at nightfall. The next day, the tendency to condensation and consequent belt of showers having extended further south and approached nearer to us, the S. S. W. wind blew *fresher* toward it, and *did not die away at nightfall*. During the evening the reflection of the lightning playing upon the tops of the thunder clouds, just visible at the north (heat-lightning, it is termed, because supposed to be unaccompanied by thunder, but in reality lightning reflected from clouds at too great a distance for the thunder to be heard), and the continuance of the southerly wind after nightfall, gave sure evidence of the coming showers the next day, and an end of the excessive heat for that time. So ended both of those long-to-be-remembered "heated terms" of 1853.

The same is probably true of the interior of the country every where. Lieutenant Maury, in the course of his investigations, and in order to ascertain the direction of the winds in the Mississippi valley during rain, addressed a number of gentlemen, and received their replies, which are published with his wind and current charts. Several answered, among other things, that, "whenever the lightning appears to linger at the north at eventide, rain almost invariably follows speedily; not so in the south." Thus it frequently is with us. If, during a hot, dry time, of a few days continuance, the lightning so lingers in the evening, and the wind continues to blow *fresh* from the southward *after nightfall*, showers will generally follow within forty-eight hours, most commonly the next day, and a cool N. N. W. or N. W. wind with a favorable change ensue. Such, at least, has been the result of my observation for many years.

Indeed this seems to be the general law in summer in the Mississippi valley, where the easterly winds are not so common as with us. To illustrate this further, I copy from a recent work by T. Bassnett, entitled the "Mechanical Theory of Storms," two short extracts, showing the manner in which belts of showers extend southerly, while progressing north-eastwardly, at Ottawa. The first occurred in August, 1853; the last, December, 1852. The first was a belt of showers; the latter would have been in August, but the lateness of the season changed its character somewhat, though not entirely, to a more regular rain, especially toward the close.

> "AUGUST 6th.—Very fine and clear all day: wind from S. W.; a light breeze; 8 P.M. frequent flashes of lightning in the northern sky; 10 P.M., a *low bank of dense clouds in north*, fringed with cirri, visible during the flash of the lightning; 12 P.M., same continues.

> "7th.—very fine and clear morning; wind S. W. moderate; noon, clouds accumulating in the northern half of the sky; *wind fresher, S. W.*; 3 P.M., a clap of thunder over head, and black cumuli in west, north, and east; 4 P.M., much thunder and scattered showers; six miles west rained very heavily; 6 P.M., the heavy

clouds passing over to the south; 10 P.M., clear again in north.

"8th.—Clear all day; wind the same (S. W.); a hazy bank visible all along on *southern horizon*.

"December 21st, 1852.—Wind N. E., fine weather.

"22d.—Thick, hazy morning, wind east, much lighter in S. E. than in N. W.; 8 A.M., a clear arch in S. E. getting more to south; noon, *very black in W. N. W.*; above, a broken layer of cirro-cumulus, the sun visible sometimes through the waves; wind around to S. E., and fresher; getting thicker all day; 10 P.M., *wind south, strong*; thunder, lightning, and heavy rain all night, with strong squalls from south.

"23d.—Wind S. W., moderate, drizzly day; 10 P.M., wind west, and getting clearer."

It is obvious that the showers at the north passed east on the evening of the 6th of August; that new showers, taking the same course, originated in the north, but more southerly next day, with S. W. wind, and that they passed east, and others formed successively further south, which passed over the place of observation late in the afternoon, and that others formed south and passed east during the night and next day, visible in a bank on the southern horizon.

Later or earlier in the spring and autumn, these brisk afternoon southerly winds continuing after nightfall, indicate moderate rains from a rainy belt extending in a similar manner, without the cumuli and thunder which attend those of mid-summer. I shall recur to this class of showers and storms when we come to their classification.

Light surface winds from south-west to west are not often storm-winds, and are usually those which the trade near the earth draws after it. Sometimes the trade seems to draw the surface wind from the S. W. and W. S. W. with considerable rapidity, and some scud a little distance above the earth. When this is so, it will be found that a storm has passed to the north of us, or a belt of rains is passing north, which may or may not have sufficient southern extension to reach us. When there have been heavy storms at the south in the spring, especially if of snow, the S. W. wind which the trade draws after it, and which comes from the snowy or chilled surface, is exceedingly "raw"—that is, damp and chilly, although not thermometrically very cold. Probably every one has noticed these "*raw*" S. W. winds of spring.

Usually, when storms and showers, which have not a southern lateral extension, pass off, the trade is very near the earth, and a light S. W. wind or calm follows for a longer or shorter period. Not unfrequently, however, our N. E. storms terminate with a S. W. wind, shifting suddenly, perhaps, just at the close of the storm, during what is sometimes called a "clearing-off-shower," or, more frequently, dying gradually away as a N. E. wind, and coming out gently from the S. W., following the retreating cloud of the storm. In such cases it is said to "clear off warm."

With us the wind rarely blows from the west, except while slowly hauling from some southerly point to the N. W. It is probably otherwise east of the lakes and in some other localities to the north-west.

Occasionally, and most frequently in March, a W. to W. N. W. wind follows storms, and blows with considerable severity, with large irregular, squally masses of scud, and sometimes a gale. Such was the character of the dry gale which crossed the country, particularly Northern New York, in March, 1854, doing great damage. These westerly winds are always accompanied by a continued depression of the barometer, and peculiar, foggy, scuddy, condensation, and should be distinguished with care from the regular and peculiar N. W. wind, as they may be, by the continued depression of the barometer, and the character of the scud. They are doubtless magnetic storms.

The remaining surface wind, the N. W., the genuine Boreas of our climate, the invariable fair-weather wind, is one of great interest. It is unique and peculiar. It is not the left-hand wind of a rotary gale, and has no immediate connection with the storm. I have known it blow moderately, fifteen successive days in winter; rising about ten A.M., and dying away at nightfall. Occasionally, but very rarely indeed, a light wind exists from the N. W. during a storm, owing probably to a focus of intensity in relation to some surface the storm covers, like the focus which exhibits itself as a clearing-off shower near the close of a storm; but the real fair-weather Boreas is a different affair altogether. Let us observe with care its peculiarities; they are instructive.

1st. It rarely blows with any considerable force beneath the trade while there are storm clouds, or any considerable condensation in it. It does not interfere with that reciprocal action which takes place between the trade and the earth, during approaching or existing storms. I have frequently seen it with its peculiar scud clouds in the N. W., waiting for the storm condensation of the trade to pass by, that full of positive electricity it might commence its sports; rushing and eddying along the surface, licking up the warm, south polar, electric rain, which stood in pools upon the ground, or rose in steamy vapor from the surface, and with its cool breath dry up the muddy roads as no degree of heat can dry them.

The annexed figure (14) shows the appearance of the northern edge of a stratus storm cloud, passing off E. N. E. at the close of the storm, which was *"clearing off from the north-west."* It is from a daguerreotype view, looking W. N. W., taken at eight o'clock in the morning, in the fall of the year. Near the horizon maybe seen the N. W. scud, forming in the N. W. wind, which is about to follow the retreating edge of the storm cloud.

Figure 15 is from a daguerreotype view, taken at eleven o'clock the same day, when the storm cloud had passed off and its edge remained visible only south of the zenith, and the north-east scud had risen up and covered the northern half of the sky, and the wind was blowing a gale from that quarter.

FIG. 14.

Another view was taken about two P.M. of the same day, when the scud had a very dark, gloomy appearance—as *dark* and *gloomy* as those of a Mexican norther—too dark to represent by a cut.

Not unfrequently in a moist summer season, after a day of showers or rain, which have had an extending formation or lateral extension from north to south, it will commence blowing in the morning, and encourage the hay-maker with the hope of fine weather. But often before noon, the milky stratus condensation above with cumuli below, will appear in the trade; the N. W. wind die away and variable airs from the east or south appear, to be followed toward night by an enlargement of the cumuli and showers. It rarely, if ever, blows fresh till the storm condensation of the trade has passed; or continues to blow after that condensation reappears. When it commences blowing after a storm, and the northern edge of the storm is not over us, we may frequently see the latter low down in the S. E. passing eastward.

FIG. 15.

NORTH VIEW.

2d. Its scud are peculiar. Every one, probably, has noticed them. They are distinct, more or less disconnected, irregular, with every form between those of the easterly scud, cumulus, and stratus, according to the season. If large, with *dark under surfaces*; forming *rapidly* and as *rapidly dissolving*; rarely dropping any rain, sometimes dropping a flurry of snow, in November or March, oftener than at any other period; sailing away to the S. E., and casting a traveling shadow as they pass on over the surface of the earth. Their electricity, particularly when white, is probably always positive, as that of all whitish clouds is supposed to be.

3d. *It is emphatically a surface wind.* The incident storm winds, the N. E. and S. E., frequently *commence blowing* under the storm, toward its point of greatest intensity, *up near the line of cirro-stratus condensation*, evidenced by the running scud; or blow there with most rapidity, and so continue for hours before the whole surface atmosphere from thence to the earth becomes involved in the movement; and sometimes without being felt below at all. Not so with the N. W. wind; it *begins at the surface* and blows there with more rapidity than above; it seems to be attracted by the earth; it interposes between the earth and the trade, wedging the trade up and occupying its place. It blows under at all seasons of the year, but most readily and strongly from a surface of snow whose electricity is always positive. Hence it blows most strongly and *continuously* when snow has fallen at the north, and prevails during winter very much in proportion to the extent and continuance of the covering of snow which invests the earth in that direction. It follows after storms, and particularly warm rains, during the autumn, winter, and spring months, which have a lateral southern extension. Whether it is increased by the snow from the surface from which it blows, or is caused by the same magnetic action which causes the great fall of snow, is a question we shall consider hereafter.

4th. It does not connect or mingle with the trade current in any way, or change or divert the course of that current; but interposes between it and the earth, elevating the trade in proportion to its own volume, above the influences of the earth (when the trade becomes free from condensation, and singularly, clear); and raising *proportionately* the barometer. An experienced observer can frequently estimate, with considerable accuracy, the rise of the barometer, by measuring with his eye, (when the clouds will enable him to do so,) the depth of this interposed N. W. current. The barometer rarely rises after a storm, for twenty-four or forty-eight hours if the wind continues at any point from S. W. to W. N. W., but always rapidly as soon as the genuine N. W. current with any considerable depth interposes and elevates the trade.

It will be obvious to every one, I think, certainly, if they will hereafter study the subject and observe for themselves, that the N. W. wind does not blow away the storm; and that it follows after it, blowing over the surface which is uncovered by the storm; rarely, if ever, with any force when the body of the storm passed south of us; and that it is a purely surface wind, seemingly attracted by the peculiar magneto-electric state in which the surface of the earth is left, compared with a snow-clad surface to the north, by a recent storm, or that peculiar state of the trade which is left by the action of the storm. It seems to follow that magnetic wave which, passing from north to south, acts in its course upon the counter-trade, producing the storm, or belt of showers, and giving them their southern lateral extension, and will well repay future telegraphic investigation. Its electricity is intensely positive—that of the earth by the action of the storm as intensely negative.

5th. This N. W. wind occurs in all parts of the northern hemisphere, so far as we have data to determine, and its corresponding wind from the S. W. occurs in the southern hemisphere. It is identical with a class of the northers of the Gulf of Mexico, as a brief analysis of the character of the latter will show.

1st. The fall and winter *norther* is a dry wind without rain or falling weather—so is our N. W. wind.

2d. It is preceded by a falling barometer; S. E. scud and rain at the point where it blows, or to the eastward of it. So is ours when it blows a gale in the fall and spring months, which bear the nearest resemblance in climatic character to the periods when the northers blow. With this distinction, however, that our precedent rains either pass over us or to the southward, the direction of storms being E. N. E.; their precedent storms passing over or to the eastward of them as they move more to the northward.

3d. It is often preceded by a copious dew; so is ours—such dews often following light fall rains in our climate, and preceding N. W. wind.

4th. The most peculiar characteristic, however, is that the barometer rises rapidly and invariably while the norther prevails, and very much in proportion to its violence. The same is true of our genuine N. W. wind, and is not true *of any other wind* on this continent which I have observed or read of.

5th. While they are thus alike in these respects, they are unlike in no respect.

Mr. Redfield has traced them in *supposed* connection with storms which continue from that vicinity across the United States to the E. N. E., and endeavored to connect them with those storms, as the left-hand winds of a rotary gale. Obviously, I think, they are identical with our N. W. winds which also *follow*, indeed, but *are distinct from the storms*.

There are a class of northers in the Gulf of Mexico—the "Nortes del Muero Colorado"—sometimes occurring in the summer months, beginning at N. E., veering about and settling at N. N. W., and as they decline hauling round by the west to the southward. These winds correspond precisely with the hurricane winds of the West Indies, and are doubtless the incident winds of a storm traveling thence to the N. N. W. precisely as our N. E. or E. N. E gales are incident storm winds to the N. E. storms of our latitude.

In this connection we will look at the peculiarities of a West India hurricane.

"It is not a little remarkable," says Mr. Espy, speaking of the storms and hurricanes of the West Indies, "that all these storms, and *all others which have been traced to the West Indies*, traveled N. W. almost at right angles to the direction of the trade-wind in those latitudes, but very nearly, if not exactly, in the direction of an upper current of the air known to exist there toward the N. W." Substantially the same facts have been repeated by Mr. Redfield, and demonstrated by his able investigations, both there and in the Eastern Pacific, and are confirmed by the observations of Edwards, Lawson, and others, while residents there. It is a matter of surprise that gentlemen like Messrs. Redfield and Espy, who have certainly displayed great ability in the investigations of meteorological phenomena, should fail to recognize a more intimate relation between this upper current and the storms they were investigating, and to detect the general laws which govern both. The storms and hurricanes of the West Indies are comparatively of small diameter, and have little advance condensation. When they pass on to the south-western portion of North America and curve to the N. E., as they frequently do, they enlarge in front and at the sides, and their advance condensation, which is not dense enough to drop rain, extends in some cases from one to three hundred miles; and the storm itself, by the time it reaches the Alleghanies, may extend one thousand to fifteen hundred miles, and perhaps in certain magnetic states of the surface, and occasionally, may cover the entire portion of the continent, from north to south. Such, probably, was very nearly the extension of the storm investigated by Professor Loomis. In the West Indies, however, at the commencement, they vary from twenty to one hundred miles, or possibly more, in width.

First, they are preceded by a hot, sultry and oppressive atmosphere—*as are electric storms every where*—a peculiar electric state of the earth and adjacent air.

Second, the black clouds and lightning which indicate the approaching hurricane are seen to the S., S. E., and E. S. E., according to the season of the year, as we see them at the westward. During the rainy season, and when the storm, as is usual at that period, is small, and the S. E. trade blows more eastwardly, the wind at the Windward Islands, possibly, may set in at the north, and back round by the east as it progresses. So Colonel Reid thinks it sometimes does, at Barbadoes. But when the belt of rains is south, and the hurricane comes from the south-east, and is larger and more violent in its action, and the north-east winds prevail, the first effect is an increase of these trades. Soon, however, the wind hauls to the north and north-west, in opposition to its course, bearing the same relation to it that our east and north-east winds bear to storms in the United States; and the wind hauls around during the passage of the storm to the west, south-west, and south-east, and at the latter point it clears off. Mr. Edwards in his History of Jamaica says—and as a resident, his authority should be decisive as to this Island—*"that all hurricanes begin from the north*, veer back to the W. N. W., W., and S. S. W., and when they get around to the S. E. the foul weather breaks up."* Doubtless the same is true of the class of northers of which we are speaking on the Gulf of Mexico. *But with this class the barometer does not rise during the gale, and in proportion to its length and violence.* With the other class of N. W. winds—the northers of winter—it does.

The following description of two winter northers, copied from Colonel Reid's valuable work, will illustrate what has been said. *Precisely such changes from S. E. rains to N. W. winds, with blue sky and detached dark clouds—fair-weather N. W. scud—occur every autumn in October and November,* and the falling of the thermometer and rising of the barometer, after rain, and a change of the wind, are perfectly characteristic.

1843.	Wind.	Force.	Weather.	Bar.	Ther.	
Jan. 30.						
A.M. 4.	S. S. W.	2	b. c.	29.90	77	Off Tampico.
Noon.	South.	5	b. c. r.	29.86	76	Lat. 23° 41′ N., Long. 94° 50′ W.
P.M. 8.	South.	6	b. c. r.	29.84	76	Lat. 23° 41′ N., Long. 94° 50′ W.
Jan. 31.						
A.M. 4.	S. Easterly.	3	b. c.	29.90	74	Between 6 and 10 A.M., wind was variable.

Noon.	N. by W.	9	c. q. w.	29.96	76	Norther commenced at 10 A. M.
P.M. 8.	N. N. W.	9	c.	30.09	73	Lat. 22° 36′ N., Long. 95° 48′ W.
Feb. 1.						
A.M. 4.	N. N. W.	7	c. g.	30.29	63	Lat. 22° 9′ N., Long. 94° 50′ W.
Noon.	Westerly.	6	c.	30.03	67	
P.M. 8.	Calm.	0	c.	30.26	67	
Feb. 14.						
A.M. 4.	S. E.	3	b. c. r.	29.66	73	At Sacraficios.
Noon.	S. W.	4	b. c.	29.62		Norther comc'd at 5.30 P.M.
P.M. 8.	N. W. by N.	10	c. q. u.	29.72	65	
Feb. 15.						
A.M. 4.	N. W. by N.	10	c. q. u.	30.10	61	Gale moderated and again freshened about 8 A.M.
Noon.	N. W. by N.	10	c. g. q.	30.19	61	
P.M. 8.	N. W.	4	c. g.	30.20	65	
Feb. 16.						
A.M. 4.	N. W.	3	q.	30.18	62	
P.M. 8.	N. N. W.	2	c. g.	30.21	66	

b. indicates blue sky — c. detached clouds — r. rain
— v. visibility of objects — q. squalls — w. wet dew
— u. ugly threatening appearance — g. gloomy weather.

The exact counterpart of the first norther may be observed with us every fall. On the 30th January, with a rising thermometer and falling barometer, there was rain at midday. The night following was moist — the next day, about ten A.M., the wind came out N. W., with squalls and gloomy weather, a falling thermometer, and rising barometer.

The norther of Feb. 14th differed from the other only in regard to the time of the day when it commenced; the order of events was the same. The rain fell in the night — it cleared off early in the day, and the norther followed in the afternoon. This also is frequently the case with us, as every one may observe.

This brief notice of the surface winds of our climate would be incomplete without a description of those of the thunder-gust and tornado.

The former is exceedingly simple. The showers, which are accompanied with much wind, form suddenly in hot weather, and have a considerable advance condensation (frequently with obvious lateral internal action), extending eastwardly from the line of smooth cloud from which the rain is falling, or rather where the falling rain obscures the inequalities of the cloud. *The gust is never felt until the advancing condensation has passed over us*, when it takes the place of the gentle easterly breeze which previously set toward the shower. *The gust ceases as soon as the cloud has passed.* It is obviously the result of the inducing and attracting influence of the cloud upon the atmosphere near the surface of the earth as it passes over it. Let the reader watch attentively this advance condensation, from its eastern edge to the line of smooth cloud and falling rain, and he will understand at a glance this internal action of gust-clouds. The whole phenomena are simple and intelligible. A cloud approaching from a westerly point, dark and irregular from its eastern edge to the line of falling rain, where it appears smooth and of a light color; wind from the east blowing gently toward it, till the condensation is over us; then the gust following the cloud; then the rain, and in a few minutes the cloud, and wind, and rain have passed on to the east, and "sunshine" returns.

The tornado, as it is termed when it occurs upon land, "spout," if on the water, is sometimes of a different character, and as it undoubtedly had great influence in inducing the gyrating theory of Mr. Redfield, and the aspiratory theory of Mr. Espy, and has been cited by both in support of their respective theories, it deserves a more particular notice. There are several marked peculiarities attending it which determine its character.

1st. It occurs during a *peculiarly sultry and electric* state of the trade and surface atmosphere, and at a time when thunder showers are prevailing in and around the locality, and at every period of the year when such a state of the atmosphere exists. One recently occurred in Brandon, Ohio, in midwinter.

2d. There is always a cloud above, but very near the earth, between which and the earth the tornado forms and rages. It is usually described as a black cloud, ranging about 1000 feet or less above the earth, often with a whitish shaped cone projecting from it, and forming a connection with the earth; at intervals rising and breaking the connection, and again descending and renewing it with devastating energy. Its width at the surface varies from forty to one hundred and eighty rods — the most usual width being from sixty to ninety rods. Sometimes when still wider, they have more the character of thunder-gusts, and are brightly luminous.

3d. Two motions are usually visible, one ascending one near the earth and in the middle, and a gyratory one around the other. The latter is rarely felt, or its effects observed, near the earth. Occasionally, and at intervals, objects are thrown obliquely backward by it.

4th. It is composed, at the surface of the earth, of *two lateral currents*, a northerly and southerly one, varying in direction, but normally at right angles in most cases, although not always, with its course of progression, extending from the extreme limits of its track to the axis; which currents are most distinctly defined toward the center, and upward. These currents prostrate trees, or elevate and remove every thing in their way which is detached and movable. There does not seem to be any current in advance of these lateral ones tending toward the tornado, save in rare and excepted cases, and then owing to the make of the ground or the irregular action of the currents; nor any following, except that made by the curving of the lateral currents toward the center of the spout as it moves on, and perhaps a tendency of the air to follow and supply the place of that which has been carried upward and forward, like that of water following the stern of a vessel. The south current is always the strongest, and often a little in advance of the other, and covers the greatest area. The proportion of the two currents to each other is much the same that the S. E. trades bear to the N. E. This excess in volume and strength of the southerly current will explain the irregularities in most cases, and the fact that objects are so often *taken up and carried from the south to the north side*, and so rarely from the north and carried south of the axis. These irregularities are such as attend all violent forces, and something can be found which will favor almost any theory; but the two lateral currents appear always to be the principal actors, except, perhaps, when it widens out and assumes more the character of a straightforward gust. See a collection by Professor Loomis, American Journal of Science, vol. xliii. p. 278.

The following diagram is a section of the New Haven tornado, from Professor Olmstead's map accompanying his article in the "American Journal of Science and Art," vol. 37. p. 340.

The manner in which the main currents flow is shown by their early and unresisted effect in a cornfield, as represented by the dotted lines. The direction in which the fragments of buildings were carried by the greater power of the southerly currents is shown also. And so is this irregular action, where a part of the southerly current broke through the northerly one, and prostrated two or three trees backward on the north side of the axis.

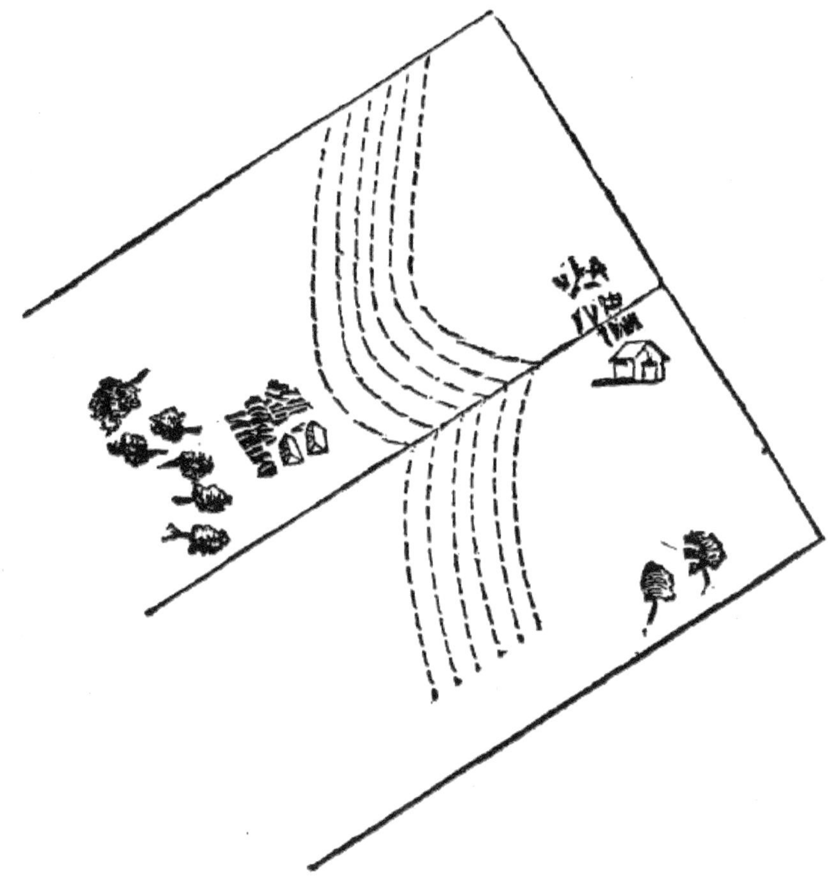

FIG. 16.

5th. This cloud, and its spout, move generally with the course of the counter-trade in the locality—*i. e.*, from some point between S. W. and W., to the eastward, but occasionally a little south of east, deflected by the magnetic wave beneath the belt of showers.

6th. Several exceedingly instructive particulars have been observed and recorded.

a. No wind is felt outside of the track, as those assert who have stood very near it, and its effects show.

b. The track is often as distinctly marked, where it passed through a wood, as if the grubbers had been there with their axes to open a path for a rail-road. The branches of the trees, projecting within its limits, are found twisted and broken off, or stripped of their leaves, while not a leaf is disturbed at the distance of a foot or two on the opposite side of the tree, and outside of the track.

c. As the spout passes over water, the latter seems to *boil up* and *rise to meet it,* and *flow up* its trunk in a *continued stream*.

d. As it passes over the land, and over buildings, fences, and other movable things, they appear to *shoot up*, instantaneously, as it were, into the air, and into fragments. If buildings are not destroyed or removed, the doors may be burst open *on the leeward side*, and gable ends *snatched out*, and roofs taken off on the *same side*, while that portion of the building which is to the windward remains unaffected.

e. Articles of clothing, and other light articles, have been carried out of buildings through open doors, or chimneys, or holes made in the roofs, and to a great distance, without *any opening* being made for the air to *blow* in.

f. If there be a discharge of electricity up the spout from the earth, like that of lightning, the intense action ceases for a time or entirely.

g. Vegetation in the track is often scorched and killed, and so of the leaves on one side of a tree, which is within the track, while those on the other side, and without the track remain unaffected. (Espy's Philosophy of Storms, 359, cited from Peltier.)

h. The active agent whatever it is, has been known to *seize hold of a chain attached to a plow* and *draw the plow about, turning the stiff sod for a considerable distance*. (See Loomis on the tornado at Stow, Ohio, American Journal of Science, vol. xxxiii. p. 368.)

i. In passing over ponds, the spout has taken up all the water and fish, and scattered them in every direction, and to a great distance.

j. The barometer falls very little during the passage of the spout. (See the Natchez hurricane of 1827, Espy page 337.) Not more than it *frequently* does during gentle showers.

k. Persons have been taken up, carried some distance, and if not projected against some object in the way, or some object against them, have usually been *set down gently and uninjured*.

l. Buildings which stood upon posts, with a free passage for the air under them, although in the path of the tornado, escaped undisturbed. (Olmstead's account of the New Haven tornado, American Journal of Science, vol. xxxvii. p 340.)

m. A chisel taken from a chest of tools, and stuck fast in the wall of the house. (Ibid.)

n. Fowls have had all their feathers stripped from them in an instant and run about naked but uninjured.[5]

o. Articles of furniture, etc., have been found torn in pieces by antagonistic forces.

p. Frames taken from looking-glasses without breaking the glass. Nails drawn from the roofs of houses without disturbing the tiles.

---

[5] All attempts to produce this result by the sudden exhaustion of air about the chickens in receivers, or shooting them from cannons, have failed, and no patent for a chicken-picker has been applied for.

q. Hinges taken from doors—*mud taken from the bed of a stream* (the water being first removed), and let down on a house covering it completely—a farmer taken up from his wagon and carried thirty rods, his horses carried an equal distance in another direction, *the harness stripped from them*, and the wagon carried off also, *one wheel not found at all.* (American Journal of Science, vol. xxxvii. p. 93.)

Pieces of timber, boards, and clapboard, driven into the side of a hill, *as no force of powder could drive them, etc., etc.*

Now to my mind, these circumstances indicate clearly, that it is not wind, *i. e.,* mere currents of air, which produces the effect, but that a *continuous current* or *stream of electricity* from the earth to the cloud exists, and carries with it from near the earth, such articles as are movable: That this stream collects from the *northerly* and *southerly* side upon the *magnetic meridian,* in *two currents* with *polarity,* which meet in their passage up at the center; curving toward the center in the posterior part as the spout moves on, when acting in a normal manner, and making the *"law of curvature"* observed: That no conceivable movement of the air alone in such limited spaces could produce such effects; or if so, that no agent but electricity could so move the air: That the air in a building could not shoot the roof upward, and into fragments; much less could the air in a cellar by any conceivable force, be made to elevate *or shoot up* the entire house, and its inmates, and contents—effects so totally unlike what takes place in gales, hurricanes, and typhoons: That elastic free air never did nor could take hold of the plow chain, and plow up the ground; or scorch and kill the vegetation; or twist the *limbs* from one side of a tree, while the most delicate leaves on the other, and within two or three feet, remained unaffected and undisturbed; or pick the chickens: That even if the expansion of the air could produce these effects—if a sudden vacuum were produced—*nothing but currents of electricity could produce the sudden vacuum,* by removing the air above.

It is well settled that atmospheric electricity can and does flow in currents with light, by experiments in relation to the brush discharge, etc. That it may do so without light or disruptive discharge, and in a stream, or as it is termed, by convection, with the force and effect seen in the tornado, is perfectly consistent with what we know of it—and it is, I think clearly evinced that such is the character of the phenomena, by the fact that a sudden powerful *disruptive* discharge, *with light, up the spout,* produces an instantaneous partial or total suspension of its action; to be renewed as the cloud passes over *another* and more highly charged *portion* of the *earth's surface.* Peltier gives instances where the spout has been entirely and instantaneously destroyed by such a sudden and powerful discharge of electricity; marking the spot where it was so destroyed by a large hole in the earth, from which the discharge issued. And in fact these tornados are often steadily luminous, and so much so, when they occur in the night, as to enable persons to read without difficulty.

The lateral inward and upward currents, are accompanied, after they meet and unite, or seem to unite, by gyratory or circular ones. How are they produced? This question can only be answered by analogy. No permanent impressions are left by the circular currents, except to a limited extent, and in occasional instances; and observation of them has been, and must necessarily be limited and uncertain.

I have witnessed one or two on a moderate scale; but owing to the suddenness of their passage, and the confusion of the objects taken up, it was difficult to determine what the circular currents were. When the southerly current is much the strongest, it appears sometimes to cross the axis, and curve round the northerly one. Perhaps this may be all the curving that really takes place, except at the posterior part of the axis, for evidence of a curving on the south of the axis is rarely, if ever seen.

Assuming, however, that the main currents unite and form one from the earth to the cloud, *induced* circular currents would be in perfect keeping with the known laws of electricity. Such currents, and with magnetic properties, are always induced by powerful currents of voltaic electricity passing through wires. And doubtless *in all cases* powerful currents of electricity *induce attendant circular currents*. This may account for the external gyration of the spout.

Or it may be that the two lateral currents of air which attend the currents of electricity, do not unite; having opposite polarity, but pass by and around each other, in connection with the circular magnetic currents. Future observation and perhaps experimental research will determine this. But it may not be accomplished by the present generation; for the belief that tornados are mere whirlwinds, produced by the action of the sun in heating the land, is adhered to, notwithstanding they cross the intense magnetic area of Ohio in mid-winter, and seems to be ineradicable.

The proportions of different winds vary in different localities. For the benefit of those who are curious, I copy a table from an able compilation by Professor Coffin, published by the Smithsonian Institute, showing the proportion of the winds at New Haven (the station nearest to me). It will be noticed that during the year the N. W. winds blow the greatest number of days; the S. W. next; the N. E. and S. E. less than either, and about equal. It may be observed that the two latter bear about the same proportion to the whole, that our number of cloudy and stormy days, averaging about ninety, bear to the whole number of days in the year.

Course.	1804.	1811.	1812.	1813.	Total.
N.	143	105	90	111	449
N. E.	99	207	138	138	582
E.	33	18	22	23	96
S. E.	131	108	135	110	484
S.	58	69	113	80	320
S. W.	224	255	153	261	893
W.	81	69	102	57	309
N. W.	329	264	345	315	1253

This work of Mr. Coffin has been brought to my notice since the foregoing pages were written. The facts embodied in it will be found to comport with what I have observed and stated. In relation to the proportionate number of days in the year during which the wind blows from the different points of the compass at the several stations it is very full and able.

But it has cardinal defects. It does not show the *main currents* of the atmosphere. It treats the surface-winds, which are incidental, as principals. The direction of the main currents is indeed shown frequently by the mean course of the surface winds, but not uniformly or intelligibly. Nor does it distinguish between the fair weather and storm winds; nor always between the trade winds during their northern transit, and the variable winds north of the trade-wind region. Hence, the deductions derived from it disclose no general system, and sustain no theory, although many very important facts appear. Some of these, Professor Coffin found it difficult to reconcile with received theories, or satisfactorily explain. For instance, he found the prevailing winds of the United States, in Louisiana and Texas, S. and S. E.; in western Arkansas, and Missouri, southerly, and in Iowa and Wisconsin, S. W., forming a curve, and evidently connected together.

Thus, alluding to the winds west of the Mississippi, and between the parallels of 36° and 60°, he says:

> "On the American continent, west of the Mississippi, there appears to be more diversity in the mean direction of the wind, yet here it is westerly at sixteen stations out of twenty, from which observations have been obtained. The most peculiar feature in this region, is the *line* of southerly winds on the western borders of Arkansas and Missouri. It seems to form a connecting link between the winds of this zone and the south-easterly ones that we find south of it; and, in some degree, to favor an idea that has been advanced, that there is a vast eddy, extending from the western shore of the Gulf of Mexico, to the eastern shore of the Atlantic; that the easterly trade-winds of the Atlantic Ocean, when they strike the American continent, veer northwardly, and then N. E., and thus recross the Atlantic, and follow down the coast of Portugal and Africa, till they complete the circuit."

This mean prevalence of the curving winds indicates the course of the western portion of the concentrated counter-trade, of which we have so fully spoken, and to which that portion owes its rains and fertility. Doubtless the curve would have been traced somewhat further west, if observations had been obtained from more westerly stations.

The idea of an eddy, to which Professor Coffin alludes, is of course unsound; that of a counter-trade, most fully confirmed; the curve corresponding with that of the regular rains and fertility as they are known to exist.

Professor Coffin is a believer in the generally-received theory of rarefaction, as the cause of all winds. His work is published by the Smithsonian Institution, and the theory is, so far forth, nationalized. But he found it very difficult to reconcile all the facts he obtained, with the theory, and, possessing a truth-loving mind, he frankly admits it. Alluding to the prevalence of N. E. winds off the coast of Africa in the summer months, as shown by certain numbered wind-roses, he says:

> "Nos. 81, 83, 86, and 91, have caused me much perplexity.

The arrows for the warmer months evidently indicate a point of rarefaction situated to the *south* or *south-west*, and yet all the observations from which they were computed were taken within a few hundred miles of the African coast and desert of Sahara; a region, the annual range of whose temperature must be exceedingly great. The only way in which I can account for a fact so astonishing, is, by supposing the deflecting forces at these numbers to be secondary to the influence which we see so strongly marked in Nos. 88, 89, and 90. Let us, then, first devote our attention to these."

(We have not space for the map of Professor Coffin, nor is it necessary to insert it. The numbers 81, 83, 86, and 91, refer to respective portions of the Atlantic, west of Africa, North of the Cape de Verdes, of 5° of latitude each, where the N. E. trades are drawing off from the coast. The Nos. 88, 89, and 90 refer to like portions *below* the Cape de Verde, where the S. W. monsoons are found under the rainy belt; and the explanation of the distinguished author is an attempt to account for the blowing of the trades *from* Sahara, by supposing them connected with the monsoons further south, which seem to blow toward it.)

"The intense heat of the Great Desert rarefies the air exceedingly from June to October, inclusive, and hence the arrows of unparalleled length (Plate XII.)," (showing the monsoon winds below the Cape de Verdes,) "pointing toward it during those months, the longest being longer than that which represents the most uniform of the trade-winds, in the ratio of 104 to 89. The influence of this rarefaction is sufficient to curve the powerful current of the trade-winds in the manner exhibited on Plate VII. Nos. 89 and 90, and to produce the not less remarkable change in No. 88, holding the current back and retarding it, so that its progressive motion in the *three* months of July, August, and September united, hardly exceeds that during any *one* of the colder months of the year. But while this is so, the trades on the western side of the Atlantic are pursuing nearly their regular track, being but slightly affected by these influences. As a consequence, the latter must leave, as it were, a partial vacuum behind them, which is filled by air flowing in from the north-east and south-east. This will account for the seeming anomaly of having a somewhat strong deflecting force directed toward mid-ocean, in the hottest part of the year, as in the numbers above referred to. *And yet it may be very naturally asked, Why does not the air from these parts supply the Great Desert directly, instead of taking a circuitous route to supply the region that supplies it? A question which, I confess, it seems difficult to answer.*"

(The italicization in the foregoing extract is mine).

Here the worthy professor finds a fact inconsistent with the theory of rarefaction—viz.: that the winds blow off shore, and toward mid-ocean, opposite Sahara, and he is "perplexed and astonished." The theory, however, must be maintained, and one of those modifying hypotheses which have made meteorology such a complicated piece of patch-work, must be invented; some "deflecting forces" found. There is the Great Desert, bordering upon the ocean, north of the Cape de Verde Islands, for a distance of six hundred miles, widening as it extends inland, whose temperature, as he says, "*must be exceedingly great;*" and doubtless is so, and yet the air, instead of blowing in upon it in a hurricane, is actually drawing off from it, and blowing towards the S. W., where the water and air do not rise above 84°. Well may he be "perplexed and astonished."

Turning south, however, to the distance of five hundred miles or more, he finds the S. W. monsoon winds, which in those months blow under the belt of rains, toward the land, in the direction of, but at a great distance from, Sahara. It is an easy matter to suppose that they reach the Great Desert and supply its vortex of rarefaction, inasmuch as they blow in a direction toward it, and distance is no impediment to supposition.

Then it is necessary to *suppose* that the S. E. and N. E. trades, at the south-west, draw so strongly to the westward as to create a partial vacuum to the S. W. of Sahara, which is filled by the winds which draw off shore, and then we have the supply brought from the distance of five hundred miles or more, by an ascending vortex, which creates a vacuum, and the air near the vortex taken away in *another* direction by a *partial* vacuum; and so an ascending *vortex*, which creates a vacuum is supplied from a distance, and a *partial vacuum* at a distance is supplied by the air near the perfect vacuum. Such an idea of a supply by a circuitous route, and secondary influence, is not very philosophical, to say the least, and Professor Coffin feels it; and to the question, Why is it so? which, he says, may very naturally be asked, he confesses there is no answer. And there would be none, even if his suppositions were based upon facts. But other questions might be asked equally difficult to be answered, viz.:

1st. Is there any rarefaction which can draw the trades to the west, and in that particular locality, in opposition to the supposed vortex of Sahara, by creating a *partial vacuum*?

2d. Are they in fact so drawn?

3d. Do the S. W. winds, south of the Cape de Verdes, and *under the rainy belt,* which in the summer months extend up to these islands, *reach the desert at all?*

These are pertinent questions, *and every one of them must be answered in the negative.* The hypothesis is without foundation, and Professor's Coffin's perplexity and astonishment must remain, until he abandons the theory of rarefaction entirely. The winds which so perplex him are nothing but the regular N. E. trades, made to originate on the coast and continent of Africa, in summer, by the northern transit of the whole machinery. They not only draw off from the desert coast, but they *blow over the desert itself* on to the ocean, and into the rainy belt upon the land, as we have already seen, and the supposed vortex of rarefaction does not exist.

That the monsoons do not reach the desert is demonstrated by the tables of Professor Coffin, and to set it at rest we will make the necessary extracts. Commencing with the region from the equator to 5° N., and from 10° to 55° W. longitude, we have the observed winds in proportion, as follows, for July and August—the south-east trades prevailing, inasmuch as the belt of rains is at this season situated further north.

LATITUDE 0° TO 5°, LONGITUDE FROM GREENWICH 10° TO 55°.

Course.	July.	August.	Course.	July.	August.
North.	0	0	S. S. W.	54	111
N. N. E.	8	2	S. W.	1	29
N. E.	6	2	W. S. W.	6	19
E. N. E.	27	16	West.	2	9
East.	31	20	W. N. W.	1	6
E. S. E.	120	96	N. W.	1	0
S. E.	216	276	N. N. W.	0	2
S. S. E.	218	443	Calm.	8	4
South.	69	279	Total	768	1,314

Here, it is evident that the S. E. trades are the prevailing winds, but their course is variable.

Ascending to the region between 5° and 10° north latitude, and 10° to 55° west longitude, the northern part of which at this season is covered by the rainy belt; we find the monsoon, the S., S. S. W., and S. W. winds, the prevailing ones in August, although the winds are variable, as usual under the rainy belt.

Course.	July.	August.	Course.	July.	August.
North.	19	6	S. S. W.	188	368
N. N. E.	26	11	S. W.	63	94
N. E.	194	32	W. S. W.	73	93
E. N. E.	30	16	West.	33	48
East.	45	29	W. N. W.	30	18
E. S. E.	36	40	N. W.	21	9
S. E.	93	53	N. N. W.	17	13
S. S. E.	225	307	Calm.	109	74
South.	239	514	Total	1,351	1,725

Ascending to the region of 10° to 15° north latitude, and 15° to 45° west longitude, we find the winds exceedingly variable, and the monsoons diminished remarkably. If Professor Coffin's theory was correct, they should increase as they approach the desert; but they in fact, diminish, and the N. E. trades are found at the north portion.

Course.	July.	August.	Course.	July.	August.
North.	17	55	S. S. W.	30	71

N. N. E.	64	74	S. W.	33	63
N. E.	155	149	W. S. W.	19	43
E. N. E.	91	71	West.	12	25
East.	83	60	W. N. W.	17	21
E. S. E.	25	26	N. W.	13	24
S. E.	17	26	N. N. W.	24	56
S. S. E.	13	33	Calm.	62	78
South.	9	44	Total	684	919

Ascending to the region between 15° and 20° north latitude, and 15° to 45° west longitude, we get north of the belt of rains *and lose the monsoons entirely although still below the desert*; and find the regular N. E. trades, with less variable winds than are found in almost any other part of the ocean.

Course.	July.	August.	Course.	July.	August.
North.	39	20	S. S. W.	0	5
N. N. E.	210	185	S. W.	0	5
N. E.	112	87	W. S. W.	8	3
E. N. E.	114	104	West.	0	1
East.	20	36	W. N. W.	0	4
E. S. E.	21	17	N. W.	3	4
S. E.	0	2	N. N. W.	3	31
S. S. E.	2	11	Calm.	20	8
South.	5	1	Total	557	526

Ascending still further to the region between 20° and 25° north latitude, and 15° and 45° west longitude, which borders, in part, on the S. W. corner of the desert, and we have not, during the month of August, a single wind between S. S. E. and W. N. W., which blows in upon the land; and *only twelve instances out of three hundred and ninety-four in this hottest month in the year, and on the southern portion of the desert, when the wind blows on shore from any quarter*. This is demonstration. The monsoon winds are confined to the rainy belt; they do not reach the desert, nor does the desert attract the winds from the ocean, or reverse, hold back, or disturb the trades.

Course.	July.	August.	Course.	July.	August.
North.	25	20	S. S. W.	3	0
N. N. E.	210	153	S. W.	2	0
N. E.	129	77	W. S. W.	13	0
E. N. E.	110	86	West.	0	0
East.	8	20	W. N. W.	0	3
E. S. E.	4	11	N. W.	2	1
S. E.	0	3	N. N. W.	5	8

| S. S. E. | 1 | 7 | Calm. | 2 | 5 |
| South. | 1 | 0 | Total | 515 | 394 |

Ascending once more, to the region between the degrees of 25 and 30, north latitude, and 15 and 45, west longitude, we find it bounded east entirely on the center of the desert. Now here, certainly, there must be evidence of the truth of the rarefaction theory, if any where on the face of the earth. Yet here, in July and August, we find the trades as regular as any where, and not more variable winds than are found in the trades toward their northern limits every where, and in August, only forty out of four hundred and twenty-nine winds, blowing directly or indirectly on shore.

Course.	July.	August.	Course.	July.	August.
North.	32	19	S. S. W.	9	6
N. N. E.	155	125	S. W.	3	9
N. E.	144	35	W. S. W.	13	14
E. N. E.	140	89	West.	12	3
East.	48	57	W. N. W.	7	7
E. S. E.	31	23	N. W.	11	1
S. E.	8	7	N. N. W.	36	6
S. S. E.	8	12	Calm.	18	12
South.	5	4	Total	680	429

It would seem to be impossible for any man to believe in the theory of rarefaction, after an examination of these tables.

Professor Coffin discovers other anomalies, for which he finds it difficult to account. Among these are the northerly tendency, in the afternoon, of the winds in Ohio, south of Lake Erie; the winds of south-western Asia, which, he says, "Are so irregular as to defy all attempts to reduce them to system;" particularizing the N. W. at Jerusalem, the westerly at Bagdad, the N. E. at Constantinople, the northerly at Trebizond, etc., etc. Jerusalem has the Mediterranean at the N. W., Bagdad has it at the west, Constantinople has the Black Sea at the N. E., Trebizond N. N. W. and N. E., and the counter-trade, as it passes over them, draws its storm-surface wind or sea-breeze, from the quarter where evaporation is greatest, and the atmosphere is most susceptible of electrical inductive influence. Precisely as it draws from the ocean and the eastward, east of the Alleghanies, from the lake region, west of the lakes, and from the northward, south of the lakes, and from the westward, east of them.

This law of attraction will explain, too, the mean prevalence of easterly winds north of the parallel of 60°, at the stations named in his work. Great Bear Lake, Great Slave Lake, and Fort Enterprise, lie east of the Rocky Mountain range which interposes between them and the Pacific, and have Hudson's Bay and other large bodies of water on the east and north. Hence, easterly winds prevail at these places. At Norway House, on Nelson's River, near the north end of Lake Winnipeg, a large body of water, which stretches off to the south, we find the south wind the prevalent one, especially in December, when the northern and north-eastern waters are frozen up, and the N. E. largely present at all seasons of the year.

At New Hernhut, in winter, when Davis' Straits are covered with floes, the prevailing wind is east, drawn from the warm, open sea east of Greenland, where the Gulf Stream is evaporating. But in June and July, when evaporation is going on over Davis' Straits and Baffin's Bay, the prevailing winds are west and south, and the east winds fall off.

Other stations are equally instructive, but I must forbear.

In relation, however, to the easterly zone of wind, of which Professor Coffin speaks, it should be added that the counter-trade, south of the magnetic pole, in high latitudes, pursues an easterly course, is near the earth, and attracts an opposite wind as it does on the east and north of the pole, in localities where the surface atmosphere is not peculiarly susceptible to its influence, and, therefore, the *winds are mainly opposite to its course.* Thus, at Melville Island, they are almost all westerly and north-westerly, for there the remnant of the counter-trade is passing west around the magnetic pole. These westerly and north-westerly winds are very light, and like the gentle easterly breeze which sets toward the cumulus clouds and summer showers.

Since most of this work was written, I have procured, and read with great pleasure, Lieutenant Maury's "Geography of the Sea." It is a work of great interest, and should be in the hands of every one. The extent of ground covered, however, made it necessary for Lieutenant Maury to introduce much matter not derived from his own investigations. In doing this, he has taken received opinions, and has thereby introduced much heresy. The view he adopts in relation to the monsoons, although the popular one with philosophers, is of that character. He says (page 222):

> "Monsoons are, for the most part, formed of trade-winds. When a trade-wind is turned back, or diverted, by over-heated districts, from its regular course at stated seasons of the year, it is regarded as a monsoon. Thus, the African monsoons of the Atlantic, the monsoons of the Gulf of Mexico, and the Central American monsoons of the Pacific, are, for the most part, formed of the north-east trade-winds, which are turned back to restore the equilibrium which the over-heated plains of Africa, Utah, Texas, and New Mexico have disturbed. When the monsoons prevail for five months at a time—for it takes about a month for them to change and become settled—then both they and the trade-winds, of which they are formed, are called monsoons."

Again (§ 476-7):

> "The agents which produce monsoons reside on the land. These winds are caused by the rarefaction of the air over large districts of country situated on the polar edge, or near the polar edge, of the trade-winds. Thus, the monsoons of the Indian Ocean are caused by the intense heat which the rays of a cloudless sun

produce, during the summer time, upon the Desert of Cobi and the burning plains of Central Asia. When the sun is north of the equator, the force of his rays, beating down upon these wide and thirsty plains, is such as to cause the vast superincumbent body of air to expand and ascend. There is, consequently, a rush of air, especially from toward the equator, to restore the equilibrium; and, in this case, the force which tends to draw the north-east trade-winds back becomes greater than the force which is acting to propel them forward. Consequently, they obey the stronger power, turn back, and become the famous south-west monsoons of the Indian Ocean, which blow from May to September inclusive.

"Of course, the vast plains of Asia are not brought up to monsoon heat *per saltum*, or in a day. They require time both to be heated up to this point and to be cooled down again. Hence, there is a conflict for a few weeks about the change of the monsoon, when neither the trade wind nor the monsoon force has fairly lost or gained the ascendency. This debatable period amounts to about a month at each change. So that the monsoons of the Indian Ocean prevail really for about five months each way, viz.: from May to September, from the south-west, in obedience to the influence of the over-heated plains, and from November to March inclusive from the north-east, in obedience to the trade-wind force."

What the "trade-wind force" is, Lieutenant Maury tells us in another paragraph, viz.: "Calorific action of the sun and diurnal rotation of the earth"—the received calorific theory. I have already shown, I think, conclusively, that there is no expansion and ascent in the supposed region of calms, which induces, or can induce, the trades; and that, in point of fact, the air on the land is cooler under the belt of rains. But as Lieutenant Maury, whose reputation is national, adopts the theory, I shall be pardoned for copying the following table, showing the difference of temperature at two cities of India, before, after, and while the belt of inter-tropical rains is over them. It will be seen that the temperature is actually less when the belt is there, viz., in July and August, than in April and May. *This should be conclusive upon that point.*

Months.	Anjarakandy.		Calcutta.	
	Rain.	Temp.	Rain.	Temp.
	M.M.		M.M.	
January,	2,26	26°,5	0,0	18°,4
February,	2,26	27°,7	67,68	21°,5
March,	6,77	28°,4	24,82	25°,6
April,	29,33	29°,8	130,84	28°,5
May,	175,96	28°,6	16,24	29°,7
June,	794,05	26°,6	575,24	29°,3
July,	807,59	25°,8	338,38	28,°1

August,	572,98	26°,0	311,31	28,°3
September,	311,31	26°,4	254,91	28,°0
October,	157,91	26°,8	42,86	27°,2
November,	65,42	26°,9	20,30	23°,0
December,	29,33	26°,5	0,0	19°,2
Year,	2955,14	27°,2	1928,74	26°,4

Anjarakandy is on the Malabar coast, between 12° and 13° north latitude. Calcutta in an angle of the Bay of Bengal, at 22° 30' north latitude. The former is in and near the focus of the monsoons, and has a temperature in July (when 18 inches of rain fall), about as low as in December.

In the foregoing table from Kaemptz, the rain is in millimetres, about twenty-five of which make an inch, and the temperature is centigrade, which may be raised to Fahrenheit by adding four fifths of the quantity and also 32°—thus, if the height of the centigrade thermometer be 25°, add to this four fifths of 25°, which is 20°, and also 32°, the result is 77°. Twenty-five centigrade is therefore equal to seventy-seven Fahrenheit.

Lieutenant Maury is not, and should not be a theorist. He occupies the position, in some sort, of a national *investigator*, and, of course, of national *instructor*. Opinions which emanate from him, or which are endorsed by him, should be accurate. Sooner or later that which he has adopted in relation to the monsoons, and some others, must be abandoned. In addition to what has already been said, I wish to call his, and the reader's attention, to several other facts and considerations in relation to the monsoons, and particularly those of India.

1st. The deserts of Cobi and Bucharia, which constitute the "burning plains" of *Central* Asia, north-east of the Indian Ocean, lie between 38° and 45° of north latitude, and under the zone of extra-tropical rains. They are not wholly rainless. They partake of that saline character which affects so much of Asia and the western part of this continent. South of them, running nearly east and west, are the lofty ranges of the Himmalaya and Kuenlun Mountains, and the table lands of Thibet. To their saline character, in part, but mainly to the interposition of these mountain ranges, depriving the counter-trade of moisture, they owe their comparative sterility. *If bountifully supplied with rains, this salt would doubtless ere this have been washed to the ocean, as it has been from other countries, once as salt as they.* But they have some rain, and more or less vegetation, and are not intensely hot. They lie too far north, and are too elevated. Their temperature is not materially different from that of the western, and comparatively desert portions of our own country, and they are utterly incapable of creating a monsoon at the Indian Ocean, and especially from the long line of Malabar coast, where the south-west monsoons are found in most strength. The sterile portions of Utah, New Mexico, and Texas are alike incapable of such effect upon the atmosphere of Central America and Mexico. These monsoons commence in May, and prevail until October, and the temperature of the air where they blow ranges with considerable regularity between 76° at night, and 84° at mid-day, on the Malabar coast, and a trifle lower in Central America.

At Fort Fillmore, El Paso, New Mexico, in latitude 32°03, the mean temperature for

May	is	68°
June	"	78°, 5'
July	"	80°, 1'
August	"	83°, 8'
September	"	77°, 9'
And for the whole period,	.	77°, 1'

At <u>Santa</u> Fé, New Mexico, the mean for

May	is	66°, 9'
June	"	72°, 5'
July	"	75°, 3'
<u>August</u>	"	72°, 9'
September	"	62°, 3'
And for the whole period,		69°, 3'
Mean of the two united,		73°, 2'

The mean of Western Texas is about 2° higher than at Fort Fillmore, and of Utah not materially different; and the mean of *Central* Asia between 38° and 45° does not materially vary from them.

Now, it is perfectly evident that during May and September the temperature of Central Asia is far below that of the Indian Ocean and India, and never materially exceeds it. Central Asia is hot, "burning," if you please, compared with more elevated, fertile, or better watered territory *in the same latitude*, and so it has been characterized; but not so, compared with the Indian Ocean, or India, where the sun is vertical. During the greater part of the time, therefore, that the monsoons are in full blast, Utah, Texas, and New Mexico, and Cobi, and the burning plains of Asia, are from 5° to 10° colder than the temperature of the place where the monsoons are blowing. Would not such a fact be perfectly conclusive in any other science except theory-swathed meteorology?

2d. The theory assumes that the heated air has an ascensive force, which causes it to rise and create a vacuum, and this vacuum, by its suction, draws in the adjoining air, which immediately ascends. The adjoining air, drawn away from its locality, leaves a vacuum, and that is filled by another rush from the S. W., and so on, till the Indian Ocean is reached, and the monsoons are accounted for.

Now, look at the difficulties:

The highest temperature that can be assumed for the air over Cobi, at any time, without disregarding facts and analogy, is 100°. What is the ascensive power of an area of atmosphere of 100°? For this we have no problem or formula, although problems and formulas abound in the science. Professor Espy relied on heated air only to give the storm a *start*. His main reliance was·on the latent heat supposed to be given out during condensation, for his ascensive storm power. But over these "burning plains" there is, according to the theory, no storm or cloud, or condensation on which that supposed reliance for expansion can be placed. What, then, is the ascension force of air at 100°? *We ought to know, for we sometimes have it as high, or within two or three degrees as high, in all the eastern and middle States.*

The monsoons blow at from twenty to twenty-five miles an hour, and sometimes more. Is that the ascensive force of air at 100°? At 25 miles an hour it would be 2,200 feet; at 20 miles, 1,760 feet; and at 10 miles, 880 feet per minute.

Does any man believe that either current exists? Why, then, do we not have our hats taken off, or light objects carried up, or have a monsoon, or, at least, have the clouds running up, when we have such elevated temperatures. *Nothing of the kind occurs with us.* Our hottest days are comparatively still days; and I have seen the cumulus sailing gently to the east, horizontally, when the air was at 98°. Why should we be exempt? Is not our air the same and our heat the same?

Again, suppose we grant that the ascensive force is equal to 20 or even 10 miles an hour, will not the adjoining air hold back somewhat to avoid leaving behind an entire vacuum? or, will it all voluntarily rush in, and leave a new complete vacuum? and, if so, why the preference of vacuums by the air, and *when, where, and why*, should the *successive vacuums stop*? Nay, would not gravity fill the second vacuum from *above*, rather than from the south-west side? and will not the air incline to rush in, to some or all these successive vacuums, from some other side than south-west? or, have these deserts the power of selecting the quarter from which their vacuum shall be filled, and of delegating it to succeeding vacuums? Would it not incline to rush in from the east and west where there are no elevations, rather than from the S. W. and over the Kuenlun Mountains, the intervening ridges and valleys of Thibet, the lofty Himmalayas, the extent of India, and the Ghaut Mountains, from three to four thousand feet high, on its eastern coast? Would it not, at least, *leak in a little*, and lessen the force with which the vacuums would draw from the far-off Indian Ocean, so that the monsoon could not blow with equal force? or, if Cobi and its fellow deserts *must* and *can* draw from an *ocean*, why not from the head of the Arabian Sea, or Bay of Bengal, or the China Sea, which are nearer, or from the Japan Sea, which is still nearer, or the Yellow Sea, which is close by? Why draw only from under the central belt of rains? Nay, what shall be done with Professor Dove? In a recent article, republished in the American Journal of Science and Art, for January, 1855, he says: "A greatly diminished atmospheric pressure taking place in summer over the *whole continent* of Asia must produce an influx from all surrounding parts; and thus we have west winds in Europe, north winds in the Icy Sea, east winds on the east coast of Asia, and south winds in India. *The monsoon itself becomes, as we see, in this point of view, only a secondary phenomena.*" This looks very like *antagonism*. Who shall we believe?

Again, suppose you get one atmosphere from the whole area, raised up by the supposed ascensive force, and at the rate of twenty-five, twenty, or even ten miles an hour, and a new volume drawn in from the south-west, and *over the mountains*: will it not take a *little time* for *that* to *heat up*? Does it heat so fast as to *keep up the ascensive force* without intermission, at twenty-five, or twenty, or ten miles the hour? What says Mr. Ericsson to this? Can he not arrange with a moderate lens, to move his engine with the rays of the summer sun? Nay, Lieutenant Maury says they can not heat up *"per saltum*, or in a day."* But according to a reasonable calculation, they must heat up the air from 80°, or less, to 100°, at the rate of 2,000 feet per minute. Heating 2,000 feet in depth, in the proportion of 20° per minute, night and day, for five months, is *"per saltum"* in a minute, and 1,440 *"saltums"* per day!

And further still, the Indian Ocean, from which the monsoons are drawn to Cobi and Central Asia to the N. E., is during those months covered by the belt of calms and rains, as heretofore stated; and the S. E. trades blowing into it are attributed to the suction created by the ascent of heated air *there*. So, then, the monsoons are blowing away from under the rainy belt, from 500 to 1000 miles, to Cobi and the burning plains of Asia, while the ascensive force of that belt is such as to draw the S. E. trades toward the very spot, a distance of 1,200 or 1,500 miles, at 20 miles an hour! What must the ascensive force over Cobi, etc., be, if, as a "stronger power," it can overcome an ascensive force over the Indian Ocean sufficient to draw the S. E. trades 1,500 miles, at 20 miles an hour; and, in addition to the force necessary to resist this central suction, not only stop or hold back the N. E. trade, but reverse it and draw it back, at 20 miles an hour, as a monsoon? Must it not be, at least, double that of the belt of calms, or the "great region of expansion," as Professor Dove calls it?

Now, I am irresistibly tempted to ask whether a meteorological theory can be too absurd for credence, and whether it would not be as well to endow the deserts with ribs and lungs, and a proboscis long enough to reach the Indian Ocean, and the necessary power of inspiration and expiration? Such a theory would avoid all difficulties, conflict with no more analogies, and, in my judgment, be as much entitled to credit as the one to which meteorologists adhere.

3d. North of the Malabar coast, in the north-west of India, lies an extensive desert. West of that is Beloochistan, with its rainless deserts. Further west are the rainless deserts of Arabia, and these three, including the Persian deserts further north, cover *as much surface* as the deserts of Cobi and Bucharia—have the sun vertical in part, and nearly so over the entire surface—*are more intensely hot*, and lie within *one third of the distance* which intervenes between that desert and the Indian Ocean off the Malabar coast, with *an open sea and* no *mountains between*. Now, look at it. The north-west desert of India, and the rainless deserts of Beloochistan and Arabia *reverse no trade* and *have no monsoon*, although the Arabian Sea heads right up among them. They do not attract one from the Indian Ocean off the Malabar coast, although not more than one third of the distance off, and without such mountains and table lands intervening as separate that coast from Cobi. It is said by Lieutenant Maury that the monsoons, *"obey the stronger force."* But which is the stronger force? Cobi, not *wholly* rainless, lying north of 35°, under the zone of extra-tropical rains, with India and the Ghauts, the Himmalaya Mountains, the table lands of Thibet, and the Kuenlun Mountains between? or the deserts of India, Beloochistan, and Arabia, *wholly rainless*, and *intensely hot, near by*, and in *open view*. There can be but one answer to this question. Nothing in the way of desert barrenness, or elevated temperature, unless it be those of Sahara, can exceed the deserts about the head of the Arabian Sea and Persian Gulf. Certainly those of Cobi can not compare with them; yet the trades blow steadily over them, although more northerly there, as every where, near their northern limits, especially on land. Says Hopkins, in his atmospheric changes:

> "If any one part of the broad expanse of the continent of Asia could be heated so as to draw air from the Arabian Sea and the Indian Ocean during the summer, it would be that part which lies between Hindoostan and the Lake of Aral, including the region between the Valley of the Oxus and Persia, and the land of this part, unlike Hindoostan, is not screened from the sun by thick vapors. But what says Burnes respecting the winds of this part? Why, that about the latter end of June, though the thermometer was at 103° in the day, 'In this country a steady wind generally blows from the north.' And on the 23d of August, after having passed the Oxus—'The heat of the sand rose to 150°, and that of the atmosphere exceeded 100°, but the wind blew steadily, nor do I believe that it would be possible to traverse this tract in summer if it ceased to blow. The steady manner in which it comes from one direction is remarkable in this inland country.' Again—'The air itself was not disturbed but by the usual north wind that blows steadily in this desert.' And he has many other similar passages."

Here there is a vast tract of country south of 35° which has a temperature often of 103°, and does not reverse the trade and create a monsoon. How utterly unphilosophical, then, to attribute the monsoons to Cobi because they "obey the stronger force!" or to attribute them to it at all.

4th. The monsoons can not be *traced from* the Malabar coast *to Cobi.* They do not exist on the south-west of Cobi and near it, where they should in greatest force, and there is no connection, in fact, shown between them. They do not often extend more than twenty-five miles inland, or to the east of the Ghauts. There are no corresponding intervening monsoons crossing India to the mountains—none over the mountains and table lands—none under the northern lee of the mountains— nor, in short, on the whole track, nor any S. W. winds except such as naturally belong to the action of the curving counter-trade.

Finally, the investigations of Commodore Wilkes on Mauna Loa, a mountain upon Hawaii, more than 13,000 feet high, and the observations of Professor Wise and other aeronauts are sufficient to put this whole matter of heated lands and ascent of the atmosphere as the cause of winds, at rest. Commodore Wilkes was encamped for about *twenty days* on Pendulum Peak, in December and January 1840. Although not up to the elevation of the counter-trade in that latitude, he was above the local clouds which form over the island during the day, where the sea breezes blow in with as great strength as any where. Indeed, he was on the top of the "lofty conical mountain" to which Caleb Williams alludes in the letter to Professor Espy I have quoted, and above the spot where Professor Espy assumed that the clouds were rising with such force as to induce the strong sea breezes of that island. During this time there were two snow-storms on Mauna Loa, and they had the wind from the S. W. during the storm, as might be expected, looking at the situation of the mountain on the western side of the island. These storms moved to the N. W., and were observed at the other islands in that direction as rain.

The local clouds lay over the island every day, as they do over active volcanic islands which are very elevated, although it was the dry season. *Nothing like an ascent of the clouds or of the currents of air from the ocean was observed.* On the contrary, the clouds formed before the sea breezes set in, and the latter blew from the different sides of the island in under the clouds, and outward again, probably on the opposite side. The whole interior of the island is elevated, and its temperature low; and *there was no elevation of temperature on the high portions of the island over which the clouds formed, and toward which the winds blew, which could create an upward current.*

> "During our stay on the summit, we took much pleasure and interest in watching the various movements of the clouds; this day in particular, they attracted our attention; the whole island beneath us was covered with a dense white mass, in the center of which was the cloud of the volcano rising like an immense dome. All was motionless until the hour arrived when the sea-breeze set in from the different sides of the island; a motion was then seen in the clouds, at the opposite extremities, both of which seemed apparently moving toward the same center, in undulations, until they became quite compact, and so contracted in space as to enable us to see a well defined horizon; at the same time there was a wind from the mountain, at right angles, that was affecting the mass, and drawing it asunder in the opposite direction. The play of these masses was at times in circular orbits, as they became influenced alternately by the different forces, until the whole was passing to and from the center in every direction, assuming every variety of form, shape and motion.

"On other days clouds would approach us from the S. W. when we had a strong N. E. trade-wind blowing, coming up with cumulus front, reaching the height of about eight thousand feet, spreading horizontally, and then dissipating. At times they would be seen lying over the island in large horizontal sheets as white as the purest snow, with a sky above of the deepest azure blue that fancy can depict. I saw nothing in it approaching to blackness at any time." (Exploring Expedition, vol iv. p. 155).

Here, in the last paragraph, we have the whole truth disclosed. The N. E. trade was blowing on Mauna Loa, 13,000 feet above the sea, and the sea-breeze blew in on the *leeward side*, its moisture condensing over the volcanic island, but without rising *up the mountain*, or *through the surface-trade*, or *above 8,000 feet.*

So, too, the celebrated aeronaut, Mr. Wise, in the course of more than a hundred ascensions, some during high wind, and others during rain storms, never met with an ascending current, except in a single instance, in the body of a hail-cloud, and then there were descending currents also, the usual intestine motion of hail-cloud with its opposite polarities.

I copy a description of his passage through the clouds of a rain-storm, and his floating a long period above them; and there was no ascending current which disturbed their horizontal repose or progression. The double layer is not uncommon—condensation taking place at the connection of the upper and lower portions of the trades, with the surrounding atmosphere; or in the trade, and by *induction in the surface atmosphere* at the same time. Such instances are frequently visible, and if his ascensions had been undertaken at other times in stormy weather he would have seen more of them.

"Before I passed the limits of the borough, a parachute, containing an animal, was dropped, which descended fast and steady, and, just as it reached the earth, my ærial ship entered a dense black body of clouds. Ten minutes were consumed in penetrating this dismal ocean of rainy vapor, occasionally meeting with great chasms, ravines, and defiles, of different shades of light and darkness. When I emerged from this ocean of clouds, a new and wonderfully magnificent scene greeted my eyes. A faint sunshine shed its warmth and luster over the surface of this vast cloud sea. The balloon rose more rapidly after it got above it. Viewing it from an elevation above the surface, I discovered it to present the same shape of the earth beneath, developing mountains and valleys, corresponding to those on the earth's surface. The profile of the cloud-surface was more depressed than that on the earth, and, in the distance of the cloud-valley a magnificent sight presented itself. Pyramids and castles, rocks and reefs, icebergs and ships, towers and domes—every thing belonging to the grand and magnificent could be seen in this distant harbor; the half-obscured sun shedding his mellow light upon it, gave it a rich and dazzling luster. They were really "castles in the air," formed of the clouds. Casting my eyes upward, I was astonished in beholding another cloud-stratum, far above the lower one; it was what is commonly termed a "mackerel sky," the sun faintly shining through it. The balloon seemed to be stationary; the clouds above and below appeared to be quiescent; the air castles, in the distance, stood to their places; silence reigned supreme; it was solemnly sublime. Solitary and alone in a mansion of the skies, my very soul swelled with emotion; I had no companion to pour out my feelings to. Great God, what a scene of grandeur! Such were my thoughts; a reverence for the works of nature, an admiration indescribable. The solemn grandeur—the very stillness that surrounded me— seemed to make a sound of praise.

"This was a scene such that I never beheld one before or after exactly like it. Two perfect layers of clouds, one not a mile above the earth; the other, about a mile higher; and, between the two, a clear atmosphere, in the midst of which the balloon stood quietly in space. It was, indeed, a strange sight—a meteorological fact, which we cannot possibly see or make ourselves acquainted with, without soaring above the surface of the earth." (History and Practice of Aeronautics, p. 209).

This is graphic. Perhaps in relation to the conformity of the upper surface of the inferior layer of clouds, to the irregularities of the earth's surface, he was misled during the enthusiasm of the moment. He is certainly mistaken as to the possibility of observing these double layers from the earth; I have seen them in hundreds of instances. But in relation to the *quiescence* of the clouds for an hour, and *the entire absence of ascending currents*, he could not be mistaken.

And now, in the absence of all direct proof to sustain the hypothesis, that the heating of the land produces ascending currents, and thereby the winds, and especially the monsoons, and in view of all the adverse evidence, I put it to Lieutenant Maury, and every sincere searcher after meteorological truth, whether the theory should not be abandoned.

CHAPTER VII.

Height of the counter-trade in different latitudes—Cause of the Calms of Cancer—Influence of mountains upon the counter-trade—Reports of Herndon and Gibbon—Focus of precipitation in the extra-tropical belt north of its southern line—Evidences of this—The elevation of the counter-trade above the earth varies in the same latitude with the variations in the phenomena of the weather—Temperature of the counter-trade—Rain dust, its origin and indications— Volcanic ashes—How far they indicate its course of progression—Question whether there is an eastern progression of the body of the atmosphere above the machinery of distribution

The counter-trade of the northern hemisphere ranges at different heights in different latitudes, in the same latitude at different seasons, and also upon different days of the same season; and, like the line of perpetual snow, has its greatest elevation in the tropics, descending gradually to the surface of the ocean at the poles. At the northern limit of the N. E. trades, it does not, ordinarily, approach the earth sufficiently near for decided reciprocal action. Hence, at that point, storms do not often originate; the winds are lighter and more variable, and calms are more frequent than at any point, except at the meeting and elevation of the trades, or in the polar regions. Doubtless this state of things is increased by the feebler action of north polar magnetism, and the irregular action of the longitudinal magnetic currents, evinced by the irregular, and often, feeble action of the trades, near their extreme limits. They are not unfrequently wholly wanting, near the northern limit, for several days in succession, and calms and baffling winds are found in their place—another effect of the irregular action of terrestrial magnetism, consequent upon the ever-changing transit of central activity from south to north, and from north to south. Upon the islands, however, and continents, which have elevated mountain peaks and ridges, especially if of volcanic origin and activity, which approach more nearly the path of the counter-trade, a different state of things exists. There, showers and gusts are frequent. Thus, upon the Sandwich Island, Kauai, the most northern one, which is within the region of the N. E. trade during ten months of the year, and upon its volcanic peaks and elevated table-lands, and north-easterly from them, over the district of Waioli, rain falls in abundance during the year, while the coastlines upon other portions of the island can not be cultivated without irrigation. (See Wilkes' Exploring Expedition, vol. iv. pp. 61 and 71; and American Journal of Science and Art, for May, 1847).

A like state of things, in degree, may be found upon the Canaries, and the more elevated of the West India Islands. The Cape de Verdes are an exception, and the Christian world are quite often called upon for contributions of provisions, to save the inhabitants of these islands from starvation. They lie at the northern limit of the equatorial belt, and for a period of two months only (July and August), are supplied with rain. If, from any cause, the belt does not move as far north as usual during any season, unbroken drought and famine are sure to overtake them. The islands contain some elevated peaks, and are of volcanic origin, but not of present volcanic activity, and the counter-trades as they issue from the equatorial belt at their highest elevation, are too far above them for reciprocal, influential action. If the islands could be placed 10° further north, we should hear no more of drought or famine from them, and their quantity of rain and fertility would be not only more permanent, but much increased. Superadded to this, is the fact, that at that point the belt of rains precipitates feebly because the S. E. trade originates upon the southern part of the continent of Africa, and the N. E. mainly, upon the desert and the Barbary States—and both are sparingly supplied with moisture.

The same state of things is strikingly obvious upon continents wherever the mountains are sufficiently elevated, even within the trade-wind region. Thus, in South America, the Andean ranges are of great elevation, and spurs and table-lands extend from them a considerable distance to the eastward. There, the S. E. and N. E. trades of the Atlantic meet in very considerable volumes, and not only is the equatorial belt much wider than upon the Atlantic and Pacific, but the counter-trades are met upon the elevated peaks and mountain-ranges, and showers and storms on their eastern slopes and summits are frequent during the dry season—down even to the extra-tropical belt. I have already said that it was probable that the great elevation of the Andes diverted and turned south a portion of the N. E. counter-trade which would otherwise pass over the western coast of Peru.

The report of Lieutenant Herndon, which has come to my notice since that was written, states facts which strongly corroborate that opinion. It seems that the trades and counter-trades actually *bank up*, in their passage to the westward, against those mountains, and the true elevation of their eastern slopes can not be barometrically ascertained. (See report of the Exploration of the Amazon, p. 261). Lieutenant Herndon says:

"I was surprised to find the temperature of boiling water at Egas to be but 208° 2', the same within 2' of a degree that it was at a point one day's journey below Tingo Maria, which village is several hundred miles above the last rapids of the Huallaga river; at Santa Cruz, two days above the mouth of the Huallaga, it was 211° 2'; at Nauta, three hundred and five miles below this, it was 211° 3'; at Pebas, one hundred and seventy miles below Nauta, 211° 1'. I was so much surprised at these results that I had put the apparatus away, thinking that its indications were valueless; but I was still more surprised, upon making the experiment at Egas, to find that the temperature of the boiling water had fallen 3° below what it was at Santa Cruz, thus giving to Egas an altitude of fifteen hundred feet above that village, which is situated more than a thousand miles up stream of it. I continued my observations from Egas downward, and found a regular increase in the temperature of the boiling water until our arrival at Pará, where it was 211° 5′.

"From an after-investigation, I am led to believe that the cause of this phenomenon arises from the fact that the trade-winds are dammed up by the Andes, and that the atmosphere in those parts is, from this cause, compressed, and, consequently, heavier than it is further from the mountains, though over a less elevated portion of the earth. The discovery of this fact has led me to place little reliance in the indications of the barometer for elevation, at the eastern foot of the Andes. It is reasonable, however, to suppose that this cause would no longer operate at Egas, nearly one thousand miles below the mouth of the Huallaga."

The report of Lieutenant Gibbon, is also exceedingly instructive. Separating from Lieutenant Herndon at Tarma, upon the Andes, he pursued a southern course, along the eastern slopes of the chain from 11° 30′ south, almost to 18° south, at Ohuro, making a journey of about 7° 30′ of latitude.

A considerable portion of this journey was over eastern and less elevated portions of the Andes; but little below, however, the line of perpetual snow. Here, during the dry season, he met with frequent showers and fogs from the eastward, but left them as he descended into the plains upon the table-land. There he found the dry season more distinctly marked; but occasional irregularities were found upon the table-lands, as every where upon corresponding elevations. The S. E. trades, however, were there obvious, during the dry season, notwithstanding the irregularities. The rainy season, from December to May, he spent at Cochabamba, and at its close he traveled north down the Madeira and its tributaries, to the Amazon. Although scarcely consistent with my prescribed limits, I can not forbear making a few extracts. Thus, when on the mountains, east of Huanvelica, in the N. E. counter-trade, he says:

"Our course is to the eastward. The snow-capped mountains are in sight to the west. Temperature of a spring 48°; air 44°. Lightning flashes all around us; as the wind whirls from *north-east* to south-west, rain and snow-flakes become hail, half the size of peas. Thunder roars and echoes through the mountains; the mules hang their heads, and travel slowly; the thinly-clad aboriginal walks shivering as he drives the train ahead; the dark cumulus cloud seems to wrap itself around us."

Again, at the Bombam Post-house, in the focus of change from cirrus to cumulus, and stratus, and storm:

"The winds are very gentle, and curl the cirrus or hairy clouds in most graceful shapes about the hoary-headed Andes, in rich and delicate clusters; when the peak is concealed, all but the blue tinge below the snow, we see a natural bridal vail. An *easterly wind* lifts and turns them to dark, cumulus clouds, settled on the frosty crown, like an old man's winter cap; the physiognomical expression is that of anger. The change is accompanied by thunder, and seems to command all around to clothe themselves for storms. The cold rain comes down in *fine drops* upon us; the day grows darker, and the *clouds press close upon the earth.*"

During an excursion east of Cuzco—

"Turning from the river, we ascend a steep ridge of mountains—the eastern range at last. A heavy mist *wafts upward as the winds drive it against the side of the Andes*, so that our view is shortened to a few hundred yards. We hope the curtain will rise that we may view the productions of the tropical valley below; but the mist thickens, and the day gets dark with heavy, heaped-up black clouds; a rain-storm follows. The grasses are thrifty, and the top of the ridge covered with a thick sod. By barometer, we stand eleven thousand one hundred feet above the level of the sea."

In May following, having spent the rainy season in Cochabamba, he travels north—

"Our route from Tarma to Oruro was south. We traveled ahead of the sun. In December, when we arrived in Cochabamba, the sun had just passed us. As soon as he did so, the rains descended heavily on this side of the ridge; it was impossible to proceed. The roads were flooded, the ravines impassable, and the arrieros put off their journey until the dry season had commenced. After the sun passed the zenith of Cochabamba, and had fairly moved the rain belt after him toward the north, then we came out from under shelter, and are now walking behind the rain belt in dry weather, while the inhabitants are actively employed in tending their crops."

So on the north of the equatorial belt, along the whole line of the Andes, up to the northern boundary of the desert valley of the Gila, rain falls on the high mountain-ranges, owing to the contiguity of the counter-trade and the diversion of showers to the north, along their eastern sides.

During the survey of the boundary line between Mexico and California, etc., by the commission under Mr. Bartlett, it became necessary to find some spot where water and grass were abundant, for the head quarters of the commission. This was found, and *could only be found*, upon the Mimbres Mountains, at an old abandoned Spanish copper mine, 7,000 or 8,000 feet above the level of the sea, surrounded with peaks of still greater height. These elevated ranges were within influential distance of the counter-trade, and here snow fell in the winter, from the extra-tropical belt, and rain, in showers, in summer, at the period of the most northerly extension of the tropical belt; when fifteen miles off, in the valley, it was unbroken drought. Mr. Bartlett thus describes it in his Personal Narrative:

"We reached this district on the 2d of May. Vegetation was

then forward, though there had been no rain. But it must be remembered that during the winter there is snow, and hence a good deal of moisture in the earth when the spring opens. The months of May and June were moderately warm. On the third of July the first rain fell. It then came in torrents, accompanied by hail, and lasted three or four hours. Many of our adobe houses were deluged with water, and the mountain-sides exhibited cataracts in every direction. The Arroyo, which passes through the village, and which furnishes barely water enough for our party and the animals, became so much swollen as to render it difficult to cross; and, by the time it had received the numerous mountain torrents, which fall into it within a mile from our camp, it became impassable for wagons, or even mules. The dry gullies became rapid streams, five or six feet deep, and sometimes fifty feet or more across. On this day, a party, in coming to the copper mines, from the plain below, *where there had been no rain*, found themselves suddenly in a region overflowing with water, so that their progress was arrested, and they were obliged to wait until the flood had subsided. After this we had occasional showers, during the months of July and August."

The location of this mountain station is near the thirty-third degree of north latitude, while the northern limit of the equatorial belt, nowhere, except upon the mountain ranges and table-lands of Mexico, extends above 25°.

There, for the reason we have been considering, it does extend further north during July and August, in occasional showers, and in the vicinity of Mount Picacho, Mr. Bartlett met one of its mountain thunder-storms on the 13th of July, on his return south through Mexico, in latitude 32°, in the following year. (Personal Narrative, vol. ii. p. 285). These showers originated in strata of counter-trade, which had followed up along the eastern side of the mountains and not from strata which had crossed them and curved to the eastward, as is shown by the course of progression of the showers.

Let us look, in this connection, at a fact or two of great interest, though not directly connected with the point in hand. The southern limit of the extra-tropical belt in winter, on the Pacific coast of North America, is in the vicinity of San Diego, at about 32°. In summer, that limit is carried up above Astoria, which is in latitude 46° 11′—about 14°—yet New Mexico receives little if any rain in winter in the vicinity of Albuquerque, but does receive a limited supply of about seven inches in summer and autumn, five and a half inches of which falls in June, July, and August. Albuquerque is in latitude 35° 13′, below the southern summer limit of the extra-tropical belt, and north of the northern limit of the equatorial belt. This anomaly is explained by the extension west over northern New Mexico, of the extreme western edge of our concentrated counter-trade, by reason of its issuing further west from the equatorial belt in its northern extension in the summer months. This western edge, in curving to the east, north-east of New Mexico, covers the north-western States, Iowa, Minnesota, Wisconsin, etc., and furnishes them that great excess of summer precipitation which is a peculiarity of their climate; and its absence further east in winter, and the very great elevation of the Rocky Mountains and other ranges over which their ordinary counter-trade of that season curves, account for the absence of much precipitation and snow there, or over the valley of the Rio Grande in New Mexico, in winter.

We may now see, too, why the western coast and the Pacific region of the continent, below 45°, are so deficient in moisture. The S. E. trades, which arise from the western portion of the south Atlantic and the continent of South America, which, if it were not for the Andes chain, in their natural course, after passing the equatorial belt, would continue on to the north-west until they passed the limits of the N. E. trades, and curve in upon the western portion of our continent below 45°, and supply it bountifully with rain, are, in part, perhaps, diverted along the eastern side of those mountains to swell the volume of our counter-trade, and in part pass them, almost exhausted of their supply of moisture by their contiguous reciprocal action. Hence, too, the deficiency of precipitation at the base of the Andes, on the western side, and the peculiar and irregular character of the winds under the western lee of the Andean range. Baffling airs and bands of calms prevail on this portion of the Pacific, except where the mountains fall off, and then there is a westerly or south-westerly monsoon under the equatorial belt. Says Lieutenant Maury in his Charts, sixth edition, p. 731:

> "The passage, under canvass, from Panama to California, as at present made, is the most tedious, uncertain, and vexatious that is known to navigators.

> "My investigations have been carried far enough to show that at certain seasons of the year a vessel bound from Panama to California, must cross at least three, at some seasons four, such meetings of winds or bands of calms, before she can enter the region of the N. E. trades. Hence the tedious passage."

Such will ever be the state of things on this continent and upon the eastern Pacific, so long as the S. E. counter-trades are compelled to pass over the mountain chain of South and Central America.

Again, if we examine carefully the belt or zone of extra-tropical rains, we shall find that the focus of greatest precipitation is considerably north of its southern limit, and that, other things being equal, this focus travels north in summer, and gives to higher latitudes their needed summer rains. This is very apparent upon the north-western portion of our continent, as the following table will show:

	Lat.	Jan.	Feb.	Mar.	Apr.	May.	June.	July.	Aug.	Sept.	Oct.	Nov.	Dec.	Year.
San Diego, Cal.	32° 41'	0.3	1.7	1.1	0.9	0.5	0.0	0.0	0.2	0.0	0.1	1.5	3.4	9.6
San Francisco.	37° 48'	1.7	0.5	4.4	2.1	0.4	0.0	0.0	0.0	0.4	0.6	3.0	5.5	18.8
Cant., Far W., Cal.	39° 02'	3.3	0.6	6.4	2.2	0.9	0.0	0.0	0.0	0.3	0.1	3.5	4.6	21.9
Astoria, Oregon.	46° 11'	27.0	10.9	6.1	4.4	5.9	2.6	0.0	2.3	1.9	6.7	13.2	6.2	87.2

Puget's S'd, Ore.	47° 07'	11.8	3.9	4.7	4.1	0.8	0.6	0.5	1.3	1.6	3.6	5.9	6.1	44.8
Sitka, Russ. Am.	57° 3'	2.5	9.6	3.5	3.3	1.9	5.9	3.7	10.1	14.8	12.7	7.4	4.2	79.5

The figures are for inches and tenths of an inch of rain.

Thus, it will be seen that in January, when the southern line is at San Diego, at the south line of California, the focus of precipitation is over Oregon; and that in August and September when the southern line is carried up and over Oregon, the focus has traveled north to Sitka, and that it is always at least 10° north of the southern line of the belt upon that coast. The increased quantities of rain which fall at the focus of precipitation there, from Oregon up, are doubtless much enhanced by the equatorial oceanic current which flows over opposite that part of the continent. A like effect, precisely, is produced in Europe. The quantity of rain which falls at Bergen, in Norway, being $87^{61}/_{100}$ inches per year, more than three times the average for that continent.

The difference shown in the foregoing table, between Astoria and Puget's Sound, is owing to the fact that the latter lies in the interior and within the coast range of mountains, while Astoria is situated at the mouth of the Columbia River, with an open view of the ocean.

A like comparative increase of precipitation in northern latitudes, in summer, is found every where varying according to the local influences which operate in the particular case. Thus,

There falls in	Winter.	Spring.	Summer.	Aut'mn.	Year.
Burlington, Vt., lat. 44° 20'	5.7	7.3	11.4	9.8	33.9
Albany, N. Y., lat. 42° 39'	8.3	9.8	12.3	10.3	40.7
Minnesota, Iowa, lat. 41° 28'	7.3	12.3	17.4	11.7	48.8
St. Peters'g, Russ., lat. 59° 56'	3.89	3.20	5.70	4.71	17.51
Pekin, China, lat. 40°	.54	3.35	18.80	2.29	25.68

Pekin lies in the northern part of China, and would have a much larger fall of rain from a concentrated counter-trade, but for the numerous mountain-ranges which intersect its path in winter, but over which it passes at a greater elevation during the summer—a peculiarity from which the eastern section of this country is most remarkably and happily free.

Thus, it is obvious that the focus of precipitation in the zone of extra tropical rains, is some 8° to 12° north of its southern line, and travels with the whole machinery in its annual transit north and south.

It is a question of some difficulty, perhaps, whether this focus is increased by the increase of magnetic action at this point, for both the line of descent of the counter-trade, and the focus of magnetic action, are carried up in a like manner, and for a like cause, and, in all probability, both concur in the result.

There is exceeding wisdom in this provision for the gradual subsidence of the counter-trade, and gradual increase of magnetic intensity, and consequent gradual precipitation. On the European continent, and over western Asia, there are 50°

of latitude to be supplied with moisture by this polar belt of rains. If the focus of precipitation was at its southern border, the counter-trade would be deprived of its moisture at that point, and little would reach the more northern portions of the globe which are to be supplied by it. But the movement of the whole machinery carries up the southern line from the south boundaries of the Barbary States on to the Mediterranean and portions of southern Europe, and the focus of precipitation and of near approach of the counter-trade to the earth, being situated far north of the southern line, is carried up correspondingly, while the combination of the moisture with the atmosphere by south polar magnetism and electricity, and the gradual descent of the counter-trade, enable it to resist, to some extent, the influence of north polar magnetism and cold, and thus retain portions of its moisture for distribution in the polar regions.

The elevation of the counter-trade above the earth varies in the same latitude with the variations in the phenomena of the weather. An attentive observation of the clouds of our climate will soon satisfy any one of this, after he has become familiar with them, so as to distinguish with certainty the clouds of the trade. Its range, in this country, is from 3,000 feet, or less, to 12,000 feet above the earth, and its depth with us probably, from 6,000 to 8,000 feet. Gay-Lussac, in his scientific experimental balloon ascension, the first of *that character* ever made, except an imperfect one just previous, by himself and Biot, found it at about 12,000 feet over Paris, and about 4,000 feet in depth. It is detected by the thermometer when much elevated.

The atmosphere grows cool as it is ascended on mountains, or by balloons. The rate of cooling is ordinarily about 1° of Fahrenheit for every 300 feet. If it were not for the equatorial current, this progressive decrease of temperature would doubtless be perfectly uniform. Of Gay-Lussac's ascension, on this point it was said:

"At forty minutes after 9 o'clock, on the morning of the 15th September, 1804, the scientific voyager ascended, as before, from the garden of the repository of models. The barometer then stood at 30.66 English inches, the thermometer at 82° Fahrenheit, and the hygrometer at 57½°. The sky was unclouded, but misty.

"During the whole of this gradual ascent, he noticed, at short intervals, the state of the barometer, the thermometer, and the hygrometer. Of these observations, amounting in all to twenty-one, he has given a tabular view. We regret, however, that he has neglected to mark the times at which they were made, since the results appear to have been very materially modified by the progress of the day. It would likewise have been desirable to have compared them with a register, noted every half hour, at the Observatory. From the surface of the earth to the height of 12,125 feet, the temperature of the atmosphere decreased regularly, from 82° to 47° 3' by Fahrenheit's scale; *but afterward it increased again, and reached to 53° 6' at the altitude of 14,000 feet;* evidently owing to the influence of the warm currents of air which, as the day advanced, rose continually from the heated ground. From that point the temperature diminished, with only slight deviations from a perfect regularity. At the height of 18,636 feet the thermometer subsided to 32° 9', on the verge of congelation; but it sunk to 14° 9' at the enormous altitude of 22,912 feet above Paris, or 23,040 feet above the level of the sea, the utmost limit of the balloon's ascent."

The high range of the barometer indicated a very considerable elevation of the trade at the time Gay-Lussac made his ascension. I am not aware that it has since been found at so great an elevation, in so high a latitude, though it is undoubtedly elevated by the interposition of a large volume of N. W. air, upon some occasions, to nearly the same altitude with us.

In the extract in relation to the ascension of Gay-Lussac, we have another of the thousand hastily-adopted and absurd hypotheses connected with the caloric theory. It is obviously and utterly *impossible* that in addition to the ordinary accumulation of heat at the surface of the earth *"as the day advanced"* — that is, *during the forenoon*, warm currents should ascend, unobserved by Gay-Lussac during an ascent of 12,000 feet — not *affecting in the least* so large an intervening body of the atmosphere or his thermometer, and in such immense volumes as to increase the warmth of a stratum of 4,000 feet in depth, an average of 3° of Fahrenheit, and to the extent of 6° at the center.

Very few balloon ascensions have been made with a view to scientific and accurate observation. But other aeronauts have met the counter-trade at different altitudes, and in both clear and stormy weather.

Recently, in 1852, four ascensions were made in England, under the direction of the Kew Observatory Committee, of the British Association. I copy from the August number of the "London, Edinburg, and Dublin Magazine," for 1853, the following condensed amount of the result:

"The ascents took place on August 17th, August 26th, October 21st, and November 10th, 1852, from the Vauxhall Gardens, with Mr. C. Green's large balloon.

"The principal results of the observations may be briefly stated as follows:

"Each of the four series of observations shows that the progress of the temperature is not regular at all heights, but that at a certain height (*varying on different days*) the regular diminution becomes arrested, and for the space of about 2,000 feet the temperature remains constant, or even increases by a small amount. It afterward resumes its downward course, continuing, for the most part, to diminish regularly throughout the remainder of the height observed. There is thus, in the curves representing the progression of temperature with height, an appearance of *dislocation*, always in the same direction, but varying in amount from 7° to 12°.

"In the first two series, viz.: August 17th and 26th, this peculiar interruption of the progress of temperature is strikingly coincident with a *large* and *rapid fall* in the temperature of the *dew-point*. The same is exhibited in a less marked manner on November 10th. On October 21st a dense cloud existed at a height of about 3,000 feet; the temperature decreased uniformly from the earth up to the *lower* surface of the cloud. When a slight rise commenced, the rise continuing through the cloud, and to about 600 feet above its upper surface, when the regular descending progression was resumed. At a short distance above the cloud, the dew-point fell considerably, but the rate of diminution of temperature does not appear to have been affected in this instance in the same manner as in the other series; the phenomenon so strikingly shown in the other three cases being perhaps modified by the existence of moisture in a *condensed* or vesicular form.

"It would appear, on the whole, that about the principal plane of condensation heat is developed in the atmosphere, which has the effect of raising the temperature of the higher air above what it would have been had the rate of decrease continued uniformly from the earth upward."

These gentlemen do not adopt the absurd explanation of the French philosophers; they account for the phenomenon by supposing heat to be *developed* at that particular part of the atmosphere; but they are equally wide of the mark. They found the excess of heat there to the extent of 7° to 12°, and on days when there was no condensation, or other assignable cause for its *development*.

The temperature of the counter-trade partakes, doubtless, of the temperature of the adjoining strata at its upper and lower portion, and has never been found much, if any, higher than 60° at the center. Nor could it be expected. The trade, in its upward curving course, within the tropics, attains a considerable altitude where the atmosphere is comparatively cold, and necessarily loses a portion of its heat there, and during its northern flow. Probably its central summer range, in the latitude of Paris, is not far from 55°, and with us 60°.

The contrast between the trade and the surrounding atmosphere, in winter, is much more striking, and this has been observed particularly upon the Brocken of the Alps, and in the polar regions.

"In all seasons the temperature is higher on the Brocken, on a serene, than on a cloudy day, and, in the month of January, *the serene days were warmer than at Berlin*." (Kämtz's Meteorology, by Walker, p. 217.—Note.)

As the portion of the counter-trade, which does not become depolarized—in diminished volume—progresses toward the polar regions, it settles nearer the earth, and within the Arctic circle is found but little way above it. Thus, in December, 1821, Parry, at Winter Island, in latitude 66° 11', flew a kite, with a thermometer attached, to the height of 379 feet, and found that the temperature, instead of falling 1¼°, the usual ratio of decrease, rose ¾ of a degree.

The same thing was observed at Spitzbergen, in latitude 77° 30' north, and at Bosekop, latitude 69° 58', by a scientific commission, and by means of kites, confined balloons, and the ascent of elevations.

"In winter the temperature goes on increasing with the height, up to a certain limit, which is variable, according to the different atmospheric circumstances, the influence of which is not yet very exactly known. The hour of the day appears to be indifferent, since there exists no thermometric diurnal variation in the strata of the surface. The mean of thirty-six experiments, made with kites, or with captive balloons, at Bosekop, latitude 69° 58′ north, has given a mean rate of increase of 1° 6′ for the first hundred meters.[6] Beyond this limit, and even beyond the first 60 or 80 meters, the temperature again becomes decreasing, at first very slowly, but afterward the decrease is accelerated. The observations that have been made on the flanks, or on the summits, of mountains, during the same expeditions, entirely confirm these results. The cooling influence of a soil, that radiates its own heat for several weeks, without receiving any thing on the part of the sun, in compensation of its losses, the influence of *counter-currents from above*, coming from the west and the south-west, with a high temperature, account for this anomaly, which, in winter, represents the normal state of the most northern parts of the European continent." (Walker's Kämtz, p. 515.—Note.)

Mr. Walker is the only author, so far as I know, who has suspected the true cause of the phenomenon, viz.: "currents from above coming from the west and south-west, with a high temperature;" but the caloric theory "sticks like a burr," and he adheres also to the idea that a snow-clad surface, in the absence of the sun, can aid, by radiation, in warming the atmosphere for a distance of several hundred yards above it, increasing the warmth as the distance from the earth increases!

This contrast between the counter-trade and the adjacent atmosphere, in winter, in latitudes as low as that of the Brocken, is probably heightened by the increased warmth of the former, at that season. The S. E. trades then form under a vertical sun, and the difference of temperature can not be less than from 6° to 8°. Not unfrequently in winter and spring the rain will fall with a temperature of 50° to 55°, when the atmosphere near the earth is 10° or 20° or more, below those points; and it is frozen to every object upon which it falls. The trade stratum, from which it descends, is not warmed by "radiation" or by ascending currents from a snow-clad surface, and during a cloudy day; nor by a "development of heat" at that particular altitude, but it has brought its heat from the South Atlantic, and imparts it to the rain which forms within it. There is every reason to believe that the counter-trade flows north in a regular descending plane, not materially differing from that of the line of perpetual snow. The descent of the latter is well ascertained to be from about 16,000 feet at the equator, to *the surface* at the poles. The plane of the counter-trade is probably much the same, varying over different localities, from the varied action between it and the earth which we are considering; and probably both correspond with the increase of magnetic intensity.

---

[6] A meter is 1 yard, and .0936 of a yard.

Lieutenant Maury, in an able and original article upon the circulation of the atmosphere, conceives the bands of comparative calms at the northern limits of the trades, which he appropriately terms the *"Calms of Cancer,"* to be nodes in the circulation of the atmosphere, and that the upper or counter-trade here decends and becomes a surface wind from the S. W., as the N. E. trade is a surface wind; and that an upper current from the poles approaches and descends at the same node, to make the N. E. trade. But it is evident he adopted that conclusion too hastily, as he obviously did the conclusion that the calms of the horse latitudes were a type of all. We have seen that the latter are increased by a diversion of the counter-trade, and that they are avoided by making easting. So it may be observed that our upper current is a S. W. current, and no northerly upper current is visible, or exists over the country, however it may be in western Europe and the North Pacific, on the west of the magnetic poles, where cold, dry northerly and north-easterly winds are found. The origin and progress of storms withal demonstrates that no such node can exist.

Two points have been made in relation to the course of the counter-trade in the tropics, and are relied upon to show its progress there to the N. E., which deserve consideration.

In the first place, it is well known that "rain dust" falls in considerable quantities on the western coast of Africa, particularly about the Cape de Verde Islands, and also upon the Mediterranean and south-western Europe, where it is termed "sirocco dust."

> "This dust," says Lieutenant Maury, "when subjected to microscopic examination, is found to consist of infusoria and organisms, whose *habitat* (place of abode) is not Africa, but South America, and in the S. E. trade-wind region of South America. Professor Ehrenberg has examined specimens of sea dust, from the Cape de Verdes and the regions thereabout, from Malta, Genoa, Lyons, and the Tyrol, and he has found such a similarity among them as would not have been more striking had these specimens been all taken from the same pile.

> "South American forms he recognizes in all of them; indeed, they are the prevailing form in every specimen he has examined.

> "It may, I think, be now regarded as an established fact, that there is a perpetual upper current of air from South America to north Africa, and that the volume of air in these upper currents, which flows to the northward, is nearly equal to the volume which flows to the southward with the N. E. trade-winds, there can be no doubt," etc.

Now, it is doubtless true that this dust is transported in a counter-trade, and that such dust is found in South America, and is taken up there by sand-spouts, like those of the ocean in form and action. Both Humboldt and Gibbon have graphically described them. Yet I do not think the point well taken. South-eastward of the Cape de Verdes, where the surface-trades—which, becoming counter-trades, pass over these islands, and, recurring, pass over the Mediterranean and south-western Europe—should originate, there is a vast extent of unexplored continent in the same latitude as the portion of South America where the dust is found; and the same dry seasons, and the same spouts, in all probability, exist in both. Until it be shown that such forms have no "*habitat*" in central and southern and unexplored Africa, upon the same latitudes as in South America, it may fairly be presumed that the dust is taken up there. Indeed, the *curve* upon which this dust is found to fall, in the greatest quantities, is very remarkable, and corresponds remarkably with the *law of curvature* of the counter-trade we have considered, and with the progress of a storm upon that coast, and over the Mediterranean, investigated by Colonel Reid. (See Reid, on Storms and Variable Winds, p. 276.) This *curve clearly indicates the origin of the dust in South Africa.*

The second point is, that ashes from the volcanos of Mexico and Central America have fallen to the north-east of the place where they were ejected. Mr. Redfield has grouped these instances of volcanic eruption usually cited, and I copy from him:

> "We learn from Humboldt, that in the great eruption of Jorullo, a volcano of southern Mexico, which is 2,100 feet above the sea, in latitude 18° 45', longitude 161° 30', the roofs of the houses in Queretaro, more than 150 miles north, 37° east from the volcano, were covered with the volcanic dust. In January, 1845, an eruption took place in the volcano of Cosiguina, on the Pacific coast of Central America, in latitude 13° north, and having an elevation of 3,800 feet, the ashes from which fell on the island of Jamaica, distant 730 miles north, 60° east from the volcano. The elevated currents by which volcanic ashes are thus transported are seldom or never of a transient or fortuitous character; and these results, therefore, afford us one of the best indications of their general course. Thus, the progress of the higher portion of the trade-wind was marked by the eruption of Tuxtla, latitude 18° 30', longitude 95°, which covered the houses in Vera Cruz with ashes, at the distance of 80 miles north, 55° west, and also at Peroté, 160 miles north, 60° west. The ashes from the volcano, at St. Vincent, which fell at Barbadoes, and east of that island, in 1812, mark the course of a current from the westward, which appears there at times, in the region of clouds, and may, perhaps, be connected with the permanent winds on the Pacific coast of Mexico."

As to one of the instances cited in the foregoing paragraph, that of Tuxtla, it may be laid out of the case—the direction conforming substantially to the assumed course of the counter-trade at that point. St. Vincent lies W. N. W., or nearly so, of Barbadoes, and a N. W. or westerly surface-wind, prior to, and during storms, is common in the West Indies as the N. E. is here—both alike, blowing in opposition to the progressive course of the storm. There is nothing strange or peculiar, therefore, respecting that instance, or the existence of variable and especially S. W. currents, between the trades, with occasional partial condensation.

The falling of the ashes from Cosiguina, upon Jamaica, has long and often been cited, as proof that in the West Indies the prevailing upper currents run from the S. W. But it has been ascertained that, *during the same eruption, ashes fell 700 miles to the westward, on the deck of the Conway*, a vessel then upon the Pacific Ocean. That case, therefore, does not prove the absence of the S. E. counter-trade at the time, but only the presence of another, and a different current above or below it—and it may have been either, and transient.

So of the Jorullo instance. Investigation would probably have shown that ashes fell to the N. W., and that they were carried N. E. by a transient S. W. wind produced by the existence of a storm to the eastward, or one of those states of partial condensation of the counter-trade which often produce currents at greater distances without a storm. Not one of these cases disproves the existence of a S. E. counter-trade, and the invariable N. W. progression of the storms of those latitudes demonstrates it.

Occasional anomalous currents, depending upon storm action at considerable distance, are found in our atmosphere, and doubtless are there also. Thus, although the N. W. wind is almost invariably a surface wind, I have, in a few instances, seen a N. W. set at a considerable elevation, converging toward a peculiarly stormy state of atmosphere far south of us, about the period of the spring equinox. And so in one or two instances I think I have seen light cirro-stratus clouds *above* the counter-trade, when it ran very low, setting from the N. E., although the usual and almost invariable location of the N. E. wind is below the counter-trade and the stratus clouds of the storm. Aeronauts, too, have found these secondary currents beneath a serene and cloudless sky. Indeed, the S. E. counter-trade doubtless often induces a thin secondary current of S. W. wind between itself and the surface-trade, in the same manner that similar currents are induced with us, and every where.

A question arises here of considerable interest, which, I confess, I can not answer to my own satisfaction. It is, whether there be, or not, *an eastern progression of the body of the atmosphere above the machinery of distribution*. I have thought there was, and that in set fair weather I had seen a peculiar kind of cirro-cumulus cloud, in patches, the small cumuli very distinct and rounded, moving due east, which indicated such a current. But I am not satisfied, from my own observation, that it is so, nor is it easy to determine the question. The moisture of evaporation rarely, if ever, ascends to any considerable elevation, and the upper strata must be very dry. Hence, condensation, if it takes place, is thin, and perhaps often undiscernable. Investigations upon mountains prove little, for the winds of the inferior strata rush up their sides and over them. It is an open question, and future observation may solve it. The prevailing opinion seems to be that there is. If the theory of Oersted, in relation to the circular currents of a magnet, be true, there should be such a progression produced by opposite secondary currents, unless, indeed, it be also true that those currents are inoperative at so great a distance, or their influence barely suffices to retain the attenuated atmosphere in its place. Perhaps the investigations of Ampère conflict with it. But it is worth while, I think, for philosophers to inquire whether the transverse position of the needle upon the wire is not the effect of the central *longitudinal* currents, conforming to the circular currents of the wire, and whether it is not owing to the production of the same currents in a globe by the circular currents of Ampère, that the globe is magnetized, and the needles made to dip.

CHAPTER VIII.

Important to understand the precise character of the reciprocal action
between the earth and the counter-trade— Connection between
the width and movements of the belt of inter-tropical rains and
the volume of the trades—Its peculiarities over Africa, the Atlantic, and South America— The magnetic equator—Character of
the storms which originate in the inter-tropical belt indicate local
magnetic action—Supposed influence of volcanic action—Gulf
Stream changes its position—This the result of magnetic action—
Alternating contrasts of heat and cold, and rain and drought—
Dr. Webster's history of the weather—Spots upon the sun—Their
character and influence—Cold or warm periods during the same
decade, and during different decades —Connection between the
spots and magnetic disturbances and variations—Influence of
the moon upon the weather— No decisive inference to be drawn
from these facts, and a more critical examination necessary

It is exceedingly desirable, in a practical point of view, to understand the precise character of the reciprocal action which takes place between the earth and
the counter-trade, and produces the varied phenomena which mark our climate.
We have seen that the same laws, other things being equal, operate every where,
and that analogies may be sought in the character of those phenomena elsewhere,
under the same, or different, modifying circumstances. Looking, therefore, at the
magneto-electric movable machinery as a whole, and its influence upon the atmospheric circulation and conditions, we find many facts which point to a primary action in the counter-trade, and others that point as significantly to a primary
local-inducing-action in the earth. Let us briefly review those to which we have
alluded, and advert to some others, and see what solution of the question they
will justify:

The belt of inter-tropical rains appears to be, in width, and amount of precipitation, and annual travel north and south, proportionate to the volume of trades
which blow into it, the quantity of moisture they contain, and the elevation of the
surface over which they meet.

South America is the most thoroughly-watered country within the tropics, except, perhaps, portions of Hindoostan, Burmah, Siam, etc., on south-eastern Asia.
The contrast between both, and Africa, as far as explored, and as shown by its
rivers, is most obvious. The Amazon, alone, delivers more water to the ocean than
all the rivers of Africa.

Of the width of the belt of rains over Africa, in the interior, we know little. Its northern extension is less, by from 7° to 10°, than the same belt over South America, the West Indies, and Mexico. Probably its southern is also. Upon South America, the southern edge is carried down to Cochabamba, in latitude 18°, and probably to 25°, to the northern edge of the coast-desert of Peru, while it is rarely, if ever, found over the Atlantic below 7°, a difference of 12° to 20°. Over South America, too, the quantity of water which falls is also vastly in excess of that which falls upon the Atlantic. The main cause of these differences is obvious. The N. E. counter-trades which blow over Africa, originate on a surface which is rainless, as eastern Sahara, Egypt, Arabia, etc., or subject to a dry season by the northern ascent of the southern line of the extra-tropical belt, as the Barbary States, Syria, Persia, etc., and their supply of moisture is necessarily scanty. On the south, the S. E. trades originate, in part, upon the eastern portion of southern Africa, and, in part, upon the Indian Ocean, and from the latter source, and a portion of the Mediterranean, doubtless most of the water which falls upon Central Africa, is derived.

The N. E. and S. E. trades which blow into the inter-tropical belt upon the eastern portion of the Atlantic, originate upon similar surfaces, and with like effect. Thus, the S. E. trades, in summer, are from the Southern portion of Africa, and the N. E., in part, from the Mediterranean; and, in winter, the N. E. from the deserts, Senegambia, Nigritia, etc., and the S. E., owing to the narrowing of the African continent, mainly from the South Atlantic and Indian Oceans. Going west, the belt widens, and its range increases until the Andes are reached; but under their lee, on the western side, a totally different state of things is found, and the belt of the coast becomes broken and irregular, as we have seen in the citation from Maury.

The width, extension, and excessive precipitation of the belt, over South America, follow the same law. The South Atlantic widens out by the trending of the coast to the S. W., and furnishes a large area for the unobstructed formation and evaporative action of the S. E. trades. So the trending of the coast to the N. W., from 5° south to the northward, opens a large area for a like formation and action of the N. E. trades. No correspondingly favorable circumstances exist any where, except, perhaps, around Hindoostan, and there the fall of rain is very excessive in some places, as on the Kassaya hills, to the extent of 400 inches per annum. In addition to this, the magnetic line of no variation, and of greater intensity, which runs from our magnetic pole, obliquely, S. S. E., to its opposite and corresponding pole in the southern hemisphere, enters the Atlantic on the coast of North Carolina, and traverses it, and the eastern portion of South America, through the whole trade-wind region. The table-lands, and slopes, and high mountain peaks, meet the trades successively, as they go west, and the latter wrench from them, to an unusual extent, their moisture; depressing the line of perpetual snow, by an increase of quantity on the eastern sides, several thousand feet, as it is for a like cause depressed on the southern side of the Himmalayas. On the eastern slopes and tops of the Andes, as we have seen, and owing to their elevation, falls the moisture which, according to the working of the machinery, and the law of curvature, should bless the coast line of Peru and northern Chili, the eastern Pacific, northern Mexico, California, Utah, and New Mexico; and, while the Andes stand, the curse of comparative aridity must rest upon them all.

Southern Chili, and western Patagonia are supplied by the N. E. trades, which originate in the West Indies, the Gulf of Mexico, and the Caribbean Sea, and the Pacific, off Central America, in the neighborhood of the Bay of Panama. But there, again, the same effect of elevation is seen. The mountain slopes of southern Chili and Patagonia are abundantly supplied, and their mountain ranges are drenched with rain, while eastern Patagonia and southern Buenos Ayres, under their lee, are comparatively dry. So the S. E. trades, which originate off the western coast of South America, curve in upon, and aided by the oceanic currents, supply, abundantly, the N. W. coast of this continent, north of California; and there, too, the coast, and its elevated ranges, receive, as we have seen, a very large proportionate supply of their moisture. Substantially, the same state of things, as far as circumstances permit, is reproduced upon Malaysia, Hindoostan, etc., and the interposition of arid New Holland upon the evaporating trade-surface may be distinctly traced upon south-western Asia. Deserts abound there; the Caspian Sea receives the drainage of a very large surface, without an outlet; their southern line of extra-tropical rains is carried up very far in summer, and their dry season is intensely hot. (See an article in the American Journal of Science, for July, 1846, by Azariah Smith.)

Another fact in this connection is worthy of a moment's consideration. The magnetic equator, as sought by the dipping needle, is not coincident with the geographical one. Humboldt found it, on the Andes, at 7° 1' south, and it has been found still lower in the Atlantic. Over Africa it rises above the geographical equator, and descends again on the Indian Ocean. About midway the Pacific, it becomes coincident with the equator of the earth again. (See diagram, on page 83.) Perhaps it is not known, with certainty, why this is so. The south pole may be situated nearer the geographical pole than the north one—but this is not believed to be so, nor could it make the difference. The greatest southern depression of the magnetic equator is found where the lines of greatest intensity, and of no variation, are found; and at the more intense of these lines exists the greatest depression. From this, I think, it may be inferred that the needle is affected by the greater magnetic intensity of the northern hemisphere, to which it may yet appear the obliquity of the earth's axis is owing. However this may be, or whatever the cause, no marked effect is produced upon the trades. The S. E. trades, by reason of the greater extent of ocean-surface on which they originate, are every where the most extensive, regular, and forcible. The south polar waters, from which they rise, are every where trenching upon, and overriding, the north polar ones; and thus, by a most beneficent provision, the greater portion of the habitable surface is placed in the northern hemisphere, and the principal portion of the southern is left open to an extensive, active evaporative action, which supplies the northern habitable surface with a large excess of the needed moisture.

The condensation, and consequent precipitation, which takes place at the passing of the trades, as we have already said, over the ocean and lowlands, takes place mainly in the day-time. Upon the table-lands and mountain-ranges, it often continues during the evening and night. The morning, and early part of the day, however, in tropical countries, are generally fair at all elevations.

Storms also originate in the equatorial belt, and issuing forth in great volume and with great intensity of action, find their way up even within the Arctic circle. Those which pass over this continent, or the northern Atlantic, generally originate in the West Indies, some of them over the Caribbean Sea, some over the islands, and some over the open ocean to the east of them; and, nearly all the most violent, during the months of August, September, and October. It would seem most probable that the primary action in such cases was in the trades themselves, but it is by no means certain that such is the case. This is the class of storms of which Mr. Redfield has industriously investigated some twenty or more; Mr. Espy some, and Lieutenant Porter two. Their course, when very violent, is often more directly north than that of storms, however violent, which originate north of the calms of Cancer, owing, perhaps, to their greater paramagnetic character. This course I have myself observed, in several instances, about the period of the autumnal equinox—never, however, more southerly than from S. W. to N. E., on the parallel of 41°, except in three, and, perhaps, four, instances, when it has been S. W. by S. to N. E. by N. I know of no class of storms in relation to which the evidence of primary action in the counter-trade is stronger than in those of the class which originate on the ocean east of the Windward Islands. But it is not satisfactory as to them. Doubtless the conflict of polarities between the passing trades is sufficient to produce the showers and rains which are ordinarily found over the ocean and lowlands, in the equatorial belt; but it is doubtful whether it is sufficient to produce such extensive, long-continued, and violent action, as that which characterizes the hurricane autumnal gales.

They occur, too, at the time when the whole machinery of distribution has reversed its course, and is rapidly pursuing its journey south. It is a period of great magnetic disturbance, over both land and sea; of more active gales and local-increased precipitation. At the Magnetic Observatory of Toronto, Canada West, these disturbances are carefully and systematically observed, and their maxima, or periods of greatest disturbance occur in April and September. (See Silliman's Journal, new series, vol. xvii. p. 145.)

The tendency to volcanic action is not as great at the autumnal, as at the vernal equinox, for the reason that most of the volcanic action of the western hemisphere develops itself now upon South rather than North America. But both exist, and are active, and what are improperly termed equinoctial storms, and gales, and rains, are proverbial during, or just subsequent to, both periods with us—as they are when the same change, called the breaking up of the monsoons, takes place in the line of magnetic intensity, over southern and eastern Asia. A volume might be filled with extracts, showing, at least, most remarkable coincidences between violent volcanic action and great atmospheric disturbance. Perhaps the increased fall of rain at and after the equinoxes, in the northern hemisphere, and in certain localities subject to volcanic activity, is as strikingly illustrated by the register, kept by Mr. Johnson, on the volcanic Island of Kauai, one of the Hawaiian group, already alluded to, as in any other case, although it is by no means a singular one. The greatest fall of rain, in any month except April and October, was eight inches. In April, the fall was fourteen inches, in October, eighteen inches. Neither the equatorial, nor extra-tropical belt, were over the island during those months; but they were the N. E. trades, and the result was owing solely to the interposition of high volcanic mountains, *in a state of disturbance*, into, or near, the strata of the counter-trade. Mr. Dobson, in stating a theory to which we shall hereafter advert, advances the following proposition:

"7. Cyclones (hurricanes) begin in the immediate neighborhood of active volcanoes. The Mauritius cyclones begin near Java; the West Indian, near the volcanic series of the Caribbean Islands; those of the Bay of Bengal, near the volcanic islands on its eastern shores; the typhoons of the China Sea, near the Philippine Islands, etc."

The peculiar stormy state of the atmosphere, over the Gulf Stream, to which I have alluded, certainly affords no evidence of primary atmospheric action. It is a body of south polar water, pursuing its way under the guidance of magnetism—maintaining its polarity—arched somewhat like the roof of a house, by the outward pressure of a cold north polar current which it has met to the east of the Banks of Newfoundland, and forced to take an in-shore course to the southward, and the bodies of water which the rivers discharge, and a conflict with the north polar surface-winds which sweep over it, and fogs, and thunder, and rain, are a matter of course. Dr. Kane met a portion of this singular current in Baffin's Bay, north of 75°, which had preserved its characteristics and a considerable proportionate excess of heat, although it probably had been around Greenland, or found its way to the west, toward the magnetic pole, through some of its northern fiords or straits. (Grinnel Expedition, p. 120.)

The investigations of Lieutenant Maury show, that when the Gulf Stream turns to the eastward, crossing the lines of declination at right angles, as the counter-trades also seem to do in the same latitude, it is *carried up, in summer, several degrees to the north*, and descends again in winter—thus demonstrating its connection with the shifting magnetic machinery which controls alike the ocean, the atmosphere, and the temperature of the earth.[7]

There are other irregularities which deserve to be noticed, in this connection, although the analogical evidence they afford is far from being decisive.

I have already said that it was within my own observation, that alternating lines of heat and cold, as well as rain and drought, existed frequently, without regard to latitude, following, to some extent, the course of the counter-trade. Such lines have been observed by others.

Thus, Mr. Espy, after describing a snow-storm, which was followed by a very cold N. W. wind, of several days' continuance, says:

> "This cold air covered the whole country, from Michigan to the eastern coast of the United States, till the beginning of the great storm of the 26th January; and, what is worthy of particular notice is, that *the temperature began to increase first in the north and north-west*. On the morning of the 25th, in the north-western parts of Pennsylvania, and northern parts of New York, the *thermome-*

[7] See his map, accompanying the Geography of the Sea.

ter had already *risen in some places 30°*, and, in others, *above 40°*. While in the S. E. corner of Pennsylvania, and in the S. E. corner of New York it had not *begun to rise*. The *wind* also began to change from the *north-west* to *south* and *south-east, first* in the north-west parts of Pennsylvania and New York, some time before it commenced in the south-east of those States; and, during the whole of the 25th, the thermometer, in the north of New York, continued to rise, though the wind was blowing from the southward, where the thermometer was many degrees lower."

Thus, too, Mr. Redfield (American Journal of Science, November, 1846, p. 329):

"On the contrary, in times of the greatest depression of the thermometer, in numerous instances, the cold period has been found to have first taken effect in, or near, the tropical latitudes, and the Gulf of Mexico, and has thence been propagated toward the eastern portions of the United States, in a manner corresponding to the observed progression of storms."

This was because the cold N. W. wind which *followed* storms began to follow them as the storms curved and passed to the N. E.

They occur in Europe also. Says Kämtz:

"Such contrasts are not uncommon in Europe, and, in this respect, the Alps form a remarkable limit; for they separate the climates of the north of Europe from the Mediterranean climates, where the distribution of rain is not the same as in the center of Europe. Hence the differences between the climates of the north and south of France. *If the winter is mild in the north*, the newspapers are filled with the lamentations of the *Italians* and *Provençals* at the *severity of the cold.*"

These facts seem to indicate a primary action in the counter-trade. Probably in connection with one class of storms they do, and with another do not. I shall endeavor to show the distinction when I come to the classification of storms.

The difference of seasons in this country, and over the entire northern hemisphere, is often very great. In a remarkable work of a remarkable man—"A Brief History of Epidemic and Pestilential Diseases," by Noah Webster, published in 1799, 2 vols.—a history of the weather for about two centuries—1600 to 1799 inclusive, is given generally, and then in a tabular form. Those who think that every considerable extreme which occurs exceeds any thing before known, will do well to consult that work. Droughts are described, where "there was not a drop of rain for three or four months, and cattle were fed upon the leaves of the trees." Winters, so intensely cold that the thermometer fell to 20° below zero, at Brandywine; or so mild that there was little frost, and people upon Connecticut River plowed their fields, and the *peach trees blossomed in Pennsylvania in February*. These extremes generally existed in Europe and America at the same time, but occasionally they were opposite and alternate. Says Mr. Webster, in summing up the facts (vol. ii. p. 12): "It is to be observed that in some cases a severe winter extends to both hemispheres, sometimes to one only, and in a few cases to a part of a hemisphere only. Thus in 1607-8, 1683-4, 1762-3, 1766-7, 1779-80, 1783-4, the severity extended to both hemispheres. In 1640-41, 1739-40, and in other instances, the severe winter in Europe preceded, by one year, a similar winter in America. In a few instances, severe frost takes place in one hemisphere during a series of mild winters in the others; but this is less common. In general, the severity happens in both hemispheres at once, or in two winters, in immediate succession; and, as far as this evidence has yet appeared, this severity is closely attendant on volcanic discharges, with very few exceptions."

It will be seen that Dr. Webster (LL.D. and not M.D., and therefore the remarkable character of the work) attributes great influence to earthquakes and volcanic action. Probably he is correct in this. The present active volcanic action of the western hemisphere is nearly all within the trade-wind region, from Mexico to Peru inclusive. The West India islands are of volcanic origin, and the influence of volcanic action is not confined to a concussion of the earth, or the eruption of mud and lava. Its connection with magnetic action, and disturbance, is unquestionable. But whether they operate to increase or diminish the trades, and the extent to which they induce violent electric action and storms within and without the tropics, is a question which further observation must determine. The ripples of the ocean, compared by Lieutenant Banvard to that of a "boiling cauldron, or such as is formed by water being forced from under the gate of a mill-pond," are met with in the vicinity of volcanic islands, where hurricanes and water-spouts originate, and have been observed to precede storms, and be connected with a falling barometer. But whether they are volcanic or magneto-electric, it is difficult to determine. Dr. Webster remarks, as the result of observation, during the 17th century, that earthquakes had a N. W. and S. E. progression in the United States, and especially in New England. In a recent article, Professor Dana has examined, with great ability, the general and remarkable trending of coast lines, groups of islands, and ranges of mountains, from N. E. to S. W. and from N. W. to S. E. (American Journal of Science, May, 1847.)

The line of magnetic intensity, which connects our magnetic pole with its opposite, is now upon this continent nearly a N. W. and S. E. line, and the pole is fast traveling to the west. It may, and probably will yet, be established, that there is an intimate connection between the cause of volcanic action within the earth, to which the upheaval of the N. W. and S. E., and N. E. and S. W. ranges were due, and of magnetic action without, and between both, and the cause of *the S. E. extension* of our summer storms and belts of showers and barometric *waves*, and the *peculiar N. W. wind*. Our limits do not permit us to pursue the subject.

Much influence upon the weather has been attributed to the spots upon the sun. These spots are supposed to be breaks or openings in the luminous atmosphere or photosphere of the sun, through which its dark nucleus body is seen. Counselor Schwabe, of Dessau, has made them his study since 1826, and has arrived at some singular results. They seem to be numerous—in groups—and to appear periodically with minima and maxima of ten years. As the result of his observations, from 1826 to 1850, he gives us the following table and remarks:

Year.	Groups.	Days showing no spots.	Days of Observation.
1826	118	22	277
1827	161	2	273
1828	225	0	282
1829	199	0	244
1830	190	1	217
1831	149	3	239
1832	84	49	270
1833	33	139	267
1834	51	120	273
1835	173	18	244
1836	272	0	200
1837	333	0	168
1838	282	0	202
1839	162	0	205
1840	152	3	263
1841	102	15	283
1842	68	64	307
1843	34	149	312
1844	52	111	321
1845	114	29	332
1846	157	1	314
1847	257	0	276
1848	330	0	278
1849	238	0	285
1850	186	2	308

"I observed large spots, visible to the naked eye, in almost all the years not characterized by the minimum; the largest appeared in 1828, 1829, 1831, 1836, 1837, 1838, 1839, 1847, 1848. I regard all spots, whose diameter exceeds 50", as large, and it is only when of such a size that they begin to be visible to even the keenest unaided sight.

"The spots are, undoubtedly, closely connected with the formation of faculæ, for I have often observed faculæ, or narben, formed at the same points from whence the spots had disappeared, while new solar spots were also developed within the faculæ. Every spot is surrounded by a more or less bright, luminous cloud. I do not think that the spots exert any influence on the annual temperature. I register the height of the barometer and thermometer three times in the course of each day, but the annual mean numbers deduced from their observations have not hither-to indicated any appreciable connection between the temperature and the number of the spots. Nor, indeed, would any importance be due to the apparent indication of such a connection in individ-ual cases, unless the results were found to correspond with others derived from many different parts of the earth. If the solar spots exert any slight influence on our atmosphere, my tables would, perhaps, rather tend to show that the years which exhibit *a larger number of spots* had a *smaller number of fine days* than those exhib-iting few spots."

These observations *seem* to show that the spots exert no influence upon the weather, and to be satisfactory. But, perhaps, they are not entirely so. No effect would, of course, be expected from day to day, and perhaps the annual mean may not be seriously disturbed, and yet the spots may seriously affect the seasons. Pop-ular tradition has fixed upon certain periods, of 10, 20, and 40 years, for the return of winters of unusual severity; and the tables of Mr. Webster, and other facts, show that it is not wholly without foundation. If we, and those we have cited, are not mistaken in most of the views expressed, the natural effect of a partial intercep-tion or failure of the sun's rays, by or from the existence of the spots, would be to decrease the exciting power of the solar rays upon terrestrial magnetism, and, as a consequence, the volume of the trades and their amount of moisture. This would increase the *mean* heat of the summer in the temperate zone—for the *less* the vol-ume of trade, the less precipitation and variable wind, and succeeding polar waves of cooler air, and the greater mean heat. On the other hand, the same cause, and the feebler heating power of the sun's rays, would make the winters more severe, both from an absence of a portion of heat, derived directly from the sun's rays, and a less mitigating influence, from the action of the trade, by reason of its decreased volume. So, too, the absence of spots, and a more powerful influence from the so-lar rays, may gradually carry the machinery further north in summer, and further south in winter, and thus make the *seasons extreme* without seriously disturbing the mean of the year. And both these may occur in a more marked degree over our intense magnetic area than in Europe. I am satisfied that they do so occur. That the partial failure of the sun's rays limits the transit of the machinery, and the volume of the trades during the latter half of the decade, and extends the transit and in-creases the volume during the first half, producing an occasional severe summer drought and severe winter, in the warmest portion of the decade. And that the variations correspond with the difference in the character and number of the spots in different decades, and hence the longer and shorter periods.

Turning to the tables of Dr. Webster, we find that a general tendency to ex-treme seasons does seem to exist from the 6th to the 10th year of every decade, and especially of every alternate decade. The periods of 1707-8, 1728, 1737 and 1739, 1749-50, 1758-9, 1779-80, 1798-9, are those in which the tendency was seen most decided. These tables are very general. The thermometer was not perfected till about 1700, and did not get into general use before 1750. There were very few meteorological registers kept, or accessible to Dr. Webster. Hence he was obliged to resort to such other sources of information as were open to him, and such statements as he found are not always entirely reliable. The oldest inhabitant is apt to express himself very strongly respecting present extremes, and fail somewhat in his recollection of those which have past. Still his tables afford general and obvious evidence of the regularity of those periodic conditions.

A.D.	Summer.	Winter.
1701	hot and dry	....
1702	hot and dry	....
1703	....	....
1704	dry Europe	....
1705	....	....
1706	hot, dry Europe	....
1707	very hot	....
1708	....	very severe
1709	....	....
1710	....	....
1711	....	cold Europe
1712	wet England	....
1713	wet England	mild
1714	dry and hot	....
1715	dry	....
1716	very dry	severe
1717	....	severe
1718	hot and wet	....
1719	....	cold America
1720	dry Europe	....
1721	....	....
1722	cold, wet	....
1723	....	cold
1724	wet England	....
1725	wet England	....
1726	....	....
1727	dry, hot Amer.	....
1728	hot Amer.	severe Europe
1729	....	....
1730	....	very cold Eng.
1731	....	....

1732	....	severe Amer.
1733	dry Eng.	....
1734	....	....
1735	wet	....
1736	wet	....
1737	....	very severe Am.
1738	....	....
1739	wet England	very severe Eng.
1740	....	very severe Am.
1741	....	....
1742	....	severe Syria
1743	hot	....
1744	....	....
1745	....	....
1746	....	....
1747	hot and dry	severe
1748	dry	....
1749	very dry	....
1750	very hot	very severe
1751	wet England	severe Amer.
1752	very hot Amer.	....
1753	....	severe
1754	....	mild Amer.
1755	....	severe Europe
1756	....	severe Syria
1757	....	....
1758	hot	....
1759	....	severe
1760	....	....
1761	very dry Amer.	....
1762	very dry Amer.	severe
1763	....	....
1764	hot Europe	....
1765	hot Europe	severe Europe
1766	hot and dry Eur.	very severe
1767	....	cold
1768	hot	....
1769	hot	....

1770	wet England	....
1771	wet Am. & England	cold Europe
1772	hot America	Am., great snow
1773	....	....
1774	....	severe Europe
1775	....	....
1776	hot	severe Europe
1777	....	....
1778	hot	mild
1779	hot Eng.	very severe
1780	....	....
1781	....	....
1782	dry Amer.	....
1783	hot	very severe
1784	hot	....
1785	dry Europe	cold
1786	cool	cold
1787	cool	....
1788	rainy Amer.	cold
1789	cool spring, hot summer	severe Eur., mild Amer.
1790	....	....
1791	very hot Am.	cold
1792	....	....
1793	hot, dry Am.	mild Amer.
1794	....	severe Europe
1795	Amer., hot, rainy	....
1796	Autumn very Dry Am.	cold Amer.
1797	cool Am.	severe Amer.
1798	very hot}	{long & severe
1799	very dry Am.}	{Amer. & Eur.

Still more definite evidence is found in the meteorological tables of Dr. Holyoke and Dr. Hildreth, and an account, by Dr. Hildreth, of the seasons when the Ohio River was closed or obstructed by ice, found in Silliman's Journal, new series, vol. xiii. p. 238.

Thus, we have, from the tables of Dr. Holyoke, the following annual means, from 1786 to 1825, inclusive. I have arranged them in periods of five years. It will be seen that there are three peculiarities observable. First, a marked difference between the first and second periods of the decade, corresponding, generally, with the presence or absence of the spots. Second, a difference in the mean of the decades which may well be supposed to correspond with the difference in the number or size of the spots since a like difference is observable in number and size, and the time when they reached their maxima and minima, in the table of Schwabe. And, third, there are occasional single cold years during the warm period, and these correspond with what the tables of Dr. Webster show for both the sixteenth and seventeenth centuries. In relation to this, it should be remembered that volcanic action is a frequent and powerful disturber of the regular action of terrestrial magnetism, and that the extremes, for that reason, are frequently meridional or local and alternating; and to that cause very great extremes, and marked exceptions, may be due, notwithstanding the spots upon the sun may exert an influence in producing hot summers and cold winters toward the close of each decade. Thus, to select an instance to illustrate this and explain an anomaly: The coldest season during the whole period, embraced in the following tables, is that of 1812. This occurs during the decrease of spots, and the warm half of the decade. Turning to the table of volcanic action, and of earthquakes, found in the Report of the British Association for 1854, we find that year was remarkable for earthquakes in the United States and South America. In December, 1811, earthquakes commenced in the valley of the Mississippi, Ohio, and Arkansas, felt also at places in Tennessee, Kentucky, Missouri, Indiana, Virginia, North and South Carolina, Georgia, and Florida, though not so severely east of the Alleghanies, *which continued until 1813.* About the same time they commenced in Caraccas, and, in March, 1812, became severe over the greater portion of the northern section of South America, and in the Atlantic. No such general and continued succession of earthquakes occurred during the other periods embraced in the tables, and the mean of the following five years was very low, embracing the memorable cold summer of 1816.

Cold Period.		Warm Period.		Cold Period.		Warm Period.	
1786	48°.53	1791	48°.963	1796	48°.678	1801	50°.432
1787	47°.88	1792	48°.44	1797	48°.135	1802	50°.794
1788	47°.676	1793	50°.96	1798	49°.471	1803	50°.24
1789	47°.68	1794	50°.768	1799	48°.291	1804	48°.328
1790	46°.53	1795	50°.173	1800	49°.989	1805	50°.792
Mean of period	47°.659	Mean	49°.901	Mean	48°.910	Mean	50°.117
1806	47°.982	1811	50°.76	1816	47°.113	1821	48°.15
1807	48°.132	1812	45°.28	1817	46°.277	1822	49°.81
1808	49°.485	1813	47°.702	1818	48°.009	1823	47°.58
1809	47°.92	1814	48°.279	1819	50°.75	1824	49°.25
1810	49°.001	1815	47°.607	1820	48°.70	1825	50°.99
Mean	48°.505	Mean	47°.925	Mean	48°.169	Mean	49°.15

    The tables of Dr. Hildreth, from 1826 to 1854, inclusive, furnish, generally, evidence of a like character. There are, however, an anomaly or two which will be observed. From 1826 to 1830, the mean is high during the period when spots were at a maximum. But that maximum embraced a much less number of spots than the two succeeding ones. A contrast appears in the tables of Dr. Hildreth, during the early period, for Dr. Holyoke's register, for 1827, puts it *below the mean*, but Dr. Hildreth's one of the *highest of the half century*. In 1835 commenced a period when the spots were much more numerous, and from 1835 to 1838, inclusive, the seasons were correspondingly below the mean. From that period to 1844 a gradual and slightly irregular rise took place, excepting the year 1843, when another cold year intervened. The table of earthquakes, published by the British Association, closes with 1842, and I have not access to any others. The occurrence of such cold years, in the warm period, at intervals during the two centuries previous, and in 1812, and onward, and evidently owing to increased volcanic action beneath the western portion of the northern hemisphere, justifies the belief that the low temperature of 1843 was owing to the same cause. The following are the means from the tables of Dr. Hildreth:

1826	54°.00	1831	50°.87	1836	50°.03	1841	52°.18	1846	53°.64
1827	54°.92	1832	52°.42	1837	51°.57	1842	52°.83	1847	52°.00
1828	55°.22	1833	54°.56	1838	50°.62	1843	50°.77	1848	52°.50
1829	52°.38	1834	52°.40	1839	52°.54	1844	53°.25	1849	52°.09
1830	54°.93	1835	50°.65	1840	52°.35	1845	52°.73	1850	51°.48
Mean	54°.29	Mean	52°.18	Mean	51°.52	Mean	52°.35	Mean	52°.32

The observations of Dr. Holyoke were made at Salem, Massachusetts; those of Dr. Hildreth at Marietta, Ohio.

The following, in relation to the freezing of the Ohio River, is evidence of a different kind, but shows the same general correspondence, and particularly *the mildness of the winters when there were few spots*, and their severity from 1836 to 1838, inclusive, when the spots were most numerous:

1829.—River open all winter—some floating ice.
1830.—River closed 27th January.
1831.—Floating ice—closed 23d January—opened 20th February.
1832.—Closed in December, which was a very cold month—opened January 8, and remained open all winter.
1833.—Open all winter.
1834.—Open all winter.
1835.—Closed January 6—opened the last of the month—cold.
1836.—Closed 28th January—opened 25th February.
1837.—Closed from 8th December to 8th February. Cold year.
1838.—Closed from 13th January to 13th March. Cold year.
1839.—Closed from 6th December to 13th January.
1840.—Closed 29th December—opened 15th January.
1841.—Closed 3d January—opened 8th do.
1842.—Open all winter.
1843.—Closed 28th November—opened 5th December—open all the rest of the winter.

> 1844.—Open all winter.
> 1845.—Open all winter.
> 1846.—Closed 5th December—opened again a few days—closed again on the 26th. It is not stated how long it remained closed.
> 1847.—Open all winter.
> 1848.—Much floating ice, but not closed—heavy rains and floods.
> 1849.—Floating ice in January, but not closed.
> 1850.—Floating ice, but not closed.
> 1851.—Open all winter—a little ice.

(December in the above table, means December previous).

This is more reliable as to the winter season than the tables of annual means—although the evidence they afford, making due allowance for the exceptions, is very striking.

I shall return to this part of the subject again.

But there is other evidence of the influence of these spots. Their connection with the irregular magnetic disturbance of the earth has been distinctly traced. Colonel Sabine, President of the British Association, in his opening address, September, 1852, after reviewing the recent discoveries in magnetism, says:—

> "It is not a little remarkable that this periodical magnetic variation is found to be identical in period, and in epochs of maxima and minima, with the periodical variation in the frequency and magnitude of the *solar spots*, which M. Schwabe has established by twenty-six years of unremitting labor. From a cosmical connection of this nature, supposing it to be finally established, it would follow that the decennial period, which we measure by our magnetic instrument, is, in fact, a solar period, manifested to us, also, by the alternately increasing and decreasing frequency and magnitude of observations on the surface of the solar disc. May we not have in these phenomena the indication of a cycle, or period of *secular change in the magnetism of the sun*, affecting visibly his gaseous atmosphere or photosphere, and sensibly modifying the magnetic influence which he exercises on the surface of our earth?"—American Journal of Science, new series, vol. xiv. p. 438.

I think it may fairly be inferred, that although these spots do not occasion the "cold spells" and "hot spells," and other transient peculiarities, they do materially affect the *mean* temperature of the year, and exert an obvious influence when at their maxima; and there is a tendency to an increase of the heat and dryness of summer, and the severity of winter, at the periods named, in our excessive climate, and a well-established connection between the spots and magnetic disturbances and variations.

Popular opinion has ever attributed to the moon a controlling effect upon the changes of the weather. If it be dry, a storm is expected *when the moon changes*; or if it be wet, dry weather. Such popular opinions are usually entitled to respect, and founded in truth. But every attempt to verify *this opinion*, by careful observation and registration, has failed. Weather-tables and lunar phases, compared for nearly one hundred years, show four hundred and ninety-one new or full moons attended by a change of the weather, and five hundred and nine without. The celebrated Olbers, after *fifty years of careful observation* and comparison, decided against it. So did the more celebrated Arago, at a more recent date—summing up the result of his observations by saying—"Whatever the progress of the sciences, never will observers, who are trustworthy and careful of their reputation, venture to foretell the state of the weather." Still, the moon may influence the weather, though she may not effect changes at her syzygies or quadratures, and this subject should not be too summarily dismissed. That the moon can not effect changes at the periods named seems philosophically obvious. She changes, for the *whole earth*, within the period of twenty-four hours; yet, how varied the state of things on different portions of its surface. The equatorial belts of trades, and drought, and rains, cover from fifty to sixty degrees of its surface, and know nothing of lunar disturbance. The extra-tropical belt of rains and variable weather moves up in its season, uncovering 10°, or more, of latitude, and admitting the trades and a six months' drought over it, as in California, regardless of the moon. Under the zone of extra-tropical rains, even upon the eastern part of the continent of North America, "dry spells" and "wet spells" exist side by side; the focus of precipitation is now in one parallel, and now in another—*storms* exist *here* and *fair weather there*, on the same continent at the same time; and as the moon's rays in her northing pass round the northern hemisphere during the twenty-four hours, they, doubtless, pass from ten to thirty or more storms, of all characters and intensities, moving in opposition to her orbit—and as many larger intervening areas of fair weather, not one of which are indebted to her for their existence, or "take thought of her coming."

The storm, which originates in the tropics, pursues its curving way now N. W., then N. E., and again north, to the Arctic circle, and, perhaps, around the magnetic pole, over gulf, and continent, and ocean, *occupying one third the time of a lunation, and two changes, perhaps, in its progress,* without any perceptible or conceivable influence from her. Yet every inhabitant of mother-earth, influenced by *coincidences remembered*, and uninfluenced by *exceptions forgotten*, looks up within his limited horizon, and devoutly expects from the agency of some phase of the moon, a change for the special benefit of his *dot* upon the earth's surface. Upon how many of these countless dots is the moon at a particular phase, or relative distance from the sun, to change fair weather to foul, or foul to fair? Upon none. The storms keep on their way;—the wet spells, and the dry spells, the cold and the hot spells alternate in their time, and though the moon turns toward them in passing, her dark face, her half face, or her full orb (the gifts of the sun, which confer no power), they do not heed her. They are originated, and are continued, by a more potent agent. They are the work of an atmospheric mechanism, as *ceaseless* in its operation as *time*, as *regular* as the *seasons, as extensive as the globe.*

Indeed, it seems as if it was expressly designed by the Creator that the moon should not interfere materially with this atmospheric machinery. She is the nearest orb; her influence would be controlling and continuous; would follow her monthly path from south to north, and with changes too violent, and intervals too long;

and would interfere with the regular fundamental operation in the trade-wind region, where she is *vertical*. Aside from the attraction of gravitation, therefore, she seems to have been so created as to be incapable of exerting any influence. She is without an atmosphere; the rays which she reflects are polarized, and without chemical or magnetic power; and, if it be true that Melloni has recently detected heat in them, by the use of a lens three feet in diameter, which could not previously be effected, its quantity is exceedingly small, and incapable of influence. Doubtless, the attraction of her mass is felt upon the earth, as the tides attest; and upon the atmosphere as well as the ocean. But the atmosphere is comparatively *attenuated*, and exceedingly so at its upper surface. Her attraction, therefore, although felt, is not influential. She seemed, to Dr. Howard, to produce in her northing and southing, a lateral tide which the barometer disclosed, but owing to the attenuated character of the atmosphere, neither the sun nor moon create an easterly and westerly tide, that is observable, except with the most delicate instruments. Sabine is believed to have detected such a tide by the barometer, at St. Helena, of one four thousandth of an inch. But even this *infinitesimal influence* may prove an error upon further investigation. There is a diurnal variation of the barometer, but it is not the result of her attraction, for it is not later each day as are the tides, exists in the deepest mines as well as upon the surface, and is demonstrably connected with the *group* of *diurnal* changes produced by the action of the sun-light and heat upon the earth's magnetism.

Can the lateral tide, if there be one, affect the weather? for in the present state of science it seems entirely certain that the moon can exert an influence in no other way.

If the received idea of many, perhaps most, meteorologists, on which all wheel barometers are constructed, that a *high barometer* necessarily produces *fair weather*, and a *low one foul*, were true, she certainly might do so. But that idea can not be sustained, and there is no known certain influence exerted by the moon upon the weather, in relation to which we have any reliable practical data.

Humboldt appears to have adopted the impression of Sir W. Herschell, that the moon aids in the dispersion of the clouds. (Cosmos, vol. iv. p. 502.) But the tendency to such dispersion is always rapid during the latter part of the day and evening, when there is no storm approaching, and the full moon renders their dissolution visible, and attracts attention to them. The Greenwich observations, also, carefully examined by Professor Loomis, fail to confirm the impression of Herschell and Humboldt, and those eminent philosophers are doubtless in this mistaken.

From this general and somewhat desultory view of the general facts, which bear analogically upon the question, no decisive inference can be drawn in relation to the seat of the primary influence which produces the atmospheric changes. The preponderance is in favor of the magnetic, or magneto-electric, action of the earth. We must come back to our own country and grapple with the question at home.

CHAPTER IX.

Examination of existing theories—Calorific theory the prevailing one—Lateral overflow of Professor Dove— Absurdity of his views in relation to them—His theory of hurricanes—Its absurdity—A new theory by Mr. Dobson— Three theories advanced by meteorologists of this country— Professor Espy's theory—Mr. Bassnett's theory—Mr. Redfield's theory—Extended examination of the latter—His theory in relation to the fall of the barometer contradictory in its character—Philosophy of the barometric change—No aid to be derived from these theories

Before proceeding to do this, however, it may be well to look at some theories which have been advanced, and to a greater or less extent adopted, and at their bearing upon the question.

The calorific theory is at present the prevailing one in Europe and in this country. Meteorologists there and here refer all atmospheric conditions and phenomena to the influence of heat. The principal applications of that theory have been considered. But within the last few years the elasticity and tension of the aqueous vapor of the atmosphere have received much attention, as exerting an auxiliary or modifying influence. Professor Dove, of Berlin, who ranks perhaps as the most distinguished meteorologist of that continent, attributes barometric variations to *lateral overflows*, and, in the upper regions, resulting from the elevation of the atmosphere by expansion; and in this view meteorologists of Europe seem generally to acquiesce. In an article sent to Colonel Sabine, and recently republished in the American Journal of Science, January, 1855, in thus attempting to account for the annual variation of barometric pressure, which occurs in Europe and Asia, and, indeed, over the entire hemisphere. He says:

> "From the combined action or the variations of aqueous vapor, and of the dry air, we derive immediately the periodical variations of the whole atmospheric pressure. As the dry air and the aqueous vapor mixed with it, press in common on the barometer, so that the up-borne column of mercury consists of two parts, one borne by the dry air, the other by the aqueous vapor, we may well understand that as with increasing temperature the air expands, and by reason of its augmented volume rises higher, and *its upper portion overflows laterally*," etc.

And in another place he says:

> "From the magnitude of the variations in the northern hemisphere, and the extent of the region over which it prevails, we must infer that *at the time of diminished pressure a lateral overflow probably takes place*," etc.

Doubtless, the mean pressure of the atmosphere, in summer, in the northern hemisphere, is less than in winter, in some localities, and greater in others, and it differs in different countries of equal temperature. And this is all very intelligible. The mean of the pressure for the month is made up by *averaging* all the *elevations* and *depressions*. During a month, showing a very low mean, the barometer may, at times, attain its *highest altitude*, if the depressions below the mean are great or more frequent. The barometer is depressed during storms, and ranges high during *set fair* weather. Ordinarily, therefore, the more stormy the season the more diminished the mean pressure; and it is a mistake to look to an overflow to account for the fact. The changes in the location of the atmospheric machinery, and consequent change in the amount and severity of falling weather, and the periodic frequency and character of storms, and consequent *periodic* depressions and elevations of the barometer, explain the annual mean variations, as they do the other phenomena. But it is perfectly consistent with the calorific theory to attempt to account for these differences by another of those ever-necessary modifications, viz.: the different tension and elasticity of aqueous vapor in different countries of equal temperature; and then to *suppose* an expansion of the whole body of the atmosphere and a lateral overflow from the place where the air is expanded, on to some other, where it is not; and thus *suppose* all necessary currents in the upper regions, setting hither and yon, by the force of gravity alone. And apparently he who is best at supposition becomes the most distinguished meteorologist. Perhaps I have already said all that I ought to be pardoned for saying, in relation to the utter absurdity of attributing all meteorological phenomena to the agency of heat; but when I find such views as those which that article contains, emanating from so distinguished a man, sanctioned by the President of the British Association, and copied into the leading journal of science in this country, I can not forbear a further and a somewhat critical examination of them. There is more error of supposition and less truth in it, than in any other article regarding the science, of equal length, which has fallen under my notice.

What is the height of this expansion? The moisture of evaporation ascends, ordinarily, but a few thousand feet. The atmosphere grows regularly cooler, from the earth to the trade, and *the increased warmth that is felt at the surface extends but little way*. Currents of warm air do not ascend. The strata maintain, substantially, their relative positions; and this is a most beneficent provision. In northern latitudes of the temperate zone, all the warmth derived from a few hours' sunshine is needed at the surface; and, deplorable, indeed, would be our condition, if the atmosphere, as fast as warmed by the rays of the sun, were to hasten up, and the frigid strata descend in its place. The earth would not be habitable. All the warm air on its surface would be rising as soon as it became warmed, and the cold air above be descending, and enveloping us with the chilling strata which are ever floating within two or three miles above us. No. Infinite wisdom has ordered it otherwise. The laws of magnetism and of static-electric induction and attraction keep the strata in their places, and preserve to us the warmth which the solar rays afford or produce. The inhabitant of the valley, in a high northern latitude, in summer, can plant, and sow, and reap, at the base of the mountain whose summit penetrates the stratum of continual congelation, and up its sides, almost to the line of perpetual snow; and, as he looks upon the fruits of his labor, and up to the snow-clad peak that towers above him, can thank his Maker for placing a warm equatorial current, a perpetual barrier, between the fertility and warmth which surround him, and the cold destructive strata above; and thank Him for not creating such a state of things, as certain meteorologists insist we shall believe He has created. Again, where are the *upper regions*, from which the lateral overflow takes place? The atmosphere is differently estimated, at from thirty to forty-five miles, or more, in height. Whatever its height may be, it is exceedingly attenuated in its "upper regions."

Gay-Lussac marked the barometer at $12^{95}/_{100}$ inches at the height of 23,040 feet. Two thirds of the atmospheric density, then, is within five miles of the earth. Air, too, is *compressible*. Allowing for the latter and the attenuation, how many miles in vertical depth, of its "*upper regions*," must move from one portion to another, to depress the barometer two inches—its range sometimes in twenty-four hours—or even half an inch? Let the computation be made, and see how startling the proposition, how utterly impossible that the theory can be true.

The distinguished Professor, in the paper referred to, introduces his theory of the formation of hurricanes, and we quote—

> "If we suppose the upper portions of the air ascending over Asia and Africa to flow off laterally, and if this takes place suddenly, it will check the course of the upper or counter-current above the trade-wind, and force it to break into the lower current.
>
> "An east wind coming into a S. W. current must necessarily occasion a rotatory movement, turning in the opposite direction to the hands of a watch. A rotatory storm, moving from S. E. to N. W., in the lower current or trade, would, in this view, be the result of the encounter of two masses of air, impelled toward each other at many places in succession, the further cause of the rotation (originating primarily in this manner) being that described by me in detail in a memoir 'On the Law of Storms,' translated in the 'Scientific Memoirs,' vol. iii. art. 7. Thus, it happens that the West India hurricanes, and the Chinese typhoons occur near the lateral confines on either side of the great region of atmospheric expansion, the typhoons being probably occasioned by the direct pressure of the air from the region of the trade-winds over the Pacific, into the more expanded air of the monsoon region, and being distinct from the storms appropriately called by the Portuguese 'temporales,' which accompany the out-burst of the monsoon when the direction of the wind is reversed."

The analogy between this, and a theory of Mr. Redfield's, will be noticed further on. But I remark, in passing, that there is not a fact or inference in this paragraph which will bear examination.

1. There is no such regular S. W. wind over the surface trade, as he supposes. Doubtless, there are, occasionally, secondary S. W. currents between the counter-trade and the surface one, with partial condensation, for much of both becomes depolarized by their reciprocal action and precipitation, and these induced S. W. currents are sometimes so strong as to usurp the place of the surface-trade, and become very violent in the latter part of hurricanes; but such is not the usual course of the upper currents of the West Indies, as the progress of storms there, and observation, prove.

2. There can not be any *periods* of extensive and *sudden* expansion over Africa. If there is any place on the earth which has a more uniformly progressive temperature, either way, and is more free from *sudden* extremes, or which is more arid and destitute of aqueous vapor, and sudden aqueous expansions, than another, it is Africa. No such occasional sudden expansions are there possible.

3. Winds do not, and can not, *"encounter."* They stratify upon each other. They are produced by the action of opposite electricity, and are *connected together* in their origin and action. The atmosphere is never free from the regular and irregular currents, however invisible for the want of condensation. Aeronauts find them in the most serene days. They exist without encounter or tendency to rotation, every where, and at all times; even over the head of the distinguished Professor, whether he sleeps or is awake. We can all see them when there is condensation, and it is rarely the case that there is not some degree of it in some of them.

4. That "Great region of expansion" is a chimera. It does not exist. It is a region of *lower temperature*, and of *condensation*, instead of *expansion* of *aqueous vapor*. The trade does not rise in it, or the S. W. wind overflow from it. See the table cited page 165.

5. The hurricanes do not originate *in the surface trades*, as he supposes. They originate in the belt of rains, the supposed "region of expansion," and issue out of it; or in the counter-trade, where volcanic elevations rise far into or above the surface trade.

6. This hypothesis can not be sustained upon his own principles. The distance between Africa and the West India Islands, where most of the hurricanes originate, is from 2,500 to 3,000 miles. These gales are small when they commence, not ordinarily over one or two hundred miles in diameter, and often less. There are trades all the way over from Africa, and S. W. winds also, if they exist, as he supposes, in the West Indies. How can it happen that this lateral overflow should pass *without effect*, over 2,500 miles of S. W. wind and trade, and concentrating the overflow of a continent over one small and chosen spot of the West Indies, *pitch down* there, and there only, and crowd the S. W. wind into the trade below? This is too much for sensible men to believe.

What does Professor Dove mean by the term *impulsion*, as applied to the winds? How are they *impelled*? It is the fundamental idea of his calorific theory, that they are *drawn* by the *suction* caused by a *vacuum*, and the vacuum created by expansion and overflow above, in obedience to the law of gravity; that the S. E. trade is drawn to the great region of expansion, and the S. W. runs from it as an overflow. But if the S. W. is driven down into the plane and place of the surface-trades, how does it continue to be impelled, and why is it not then subject to the suction of the vacuum which draws the trade? Does that vacuum *select its air*, and so attract the trade, in preference to the depressed portion of the S. W. current, that the former runs around the latter to get to the vacuum, and the latter around the former to get away from it? And does the trade, when it has got around the S. W. current, instead of going to the vacuum, continue to gyrate, and the S. W. current, instead of pursuing its regular course, gyrate also about the trade, and both move off together, regardless of the vacuum of the great region of expansion, in a new direction to the N. W., in an independent, self-sustaining, cyclonic movement, increasing in power and extent, involving extended and increasing condensation, producing the most violent electrical phenomena, and thus continuing up, even to the Arctic circle? Yes, says Professor Dove. No, say all fact, all analogy, and his own principles.

7. His theory relative to the typhoons is unintelligible. If they originate near the lateral confines of the great region of atmospheric expansion, they originate in the region of the trade-winds, for the two are identical. How the direct pressure of the air from the trade-wind over the Pacific, in the more expanded air of the monsoon region, can occasion a typhoon upon any principles, passes my comprehension. If, as Lieutenant Maury supposes, the monsoons are reversed trades, then the trade-wind and monsoon region are identical. If the monsoons are found in the belt of rains, then, the trades, upon Professor Dove's principles, pass into the monsoon region by attraction or suction, without pressure. Either way the theory is undeserving of consideration.

A new theory has recently been started by Mr. Thomas Dobson, and, although it is (like all other efforts to get the *upper strata down* to produce condensation, or those below *up*, that they may be condensed), without foundation, his collection of facts is brief and interesting. I copy his article from the London, Edinburgh, and Dublin Phil. Mag., for December, 1853. It adds to the collection of facts in relation to the connection between volcanic action and storms for the seventeenth century, made by Dr. Webster:

> The following appear to be the main facts which are available as a basis for a theory which shall comprehend all the meteors in question:

> 1st. The eruption of a submarine volcano has produced water-spouts.

> "During these bursts the most vivid flashes of lightning continually issued from the densest part of the volcano, and the volumes of smoke rolled off in large masses of fleecy clouds, gradually expanding themselves before the wind in a direction nearly horizontal, and drawing up *a quantity of water-spouts*." —(Captain

Tilland's description of the upheaval of Sabrina Island in June, 1811, Phil. Trans.)

With this significant fact may be compared the following analogous ones:

"In the Aleutian Archipelago a new island was formed in 1795. It was first observed *after a storm*, at a point in the sea from which a column of smoke had been seen to rise." — (Lyell, Principles of Geology.)

"Among the Aleutian Islands a new volcanic island appeared in the midst of *a storm*, attended with flames and smoke. After the sea was calm, a boat was sent from Unalaska with twenty Russian hunters, who landed on this island on June 1st, 1814." — (Journal of Science, vol. vii.)

"On July 24th, 1848, a submarine eruption broke out between the mainland of Orkney and the island of Strousa. Amid thunder and lightning, a very dense jet black cloud was seen to rise from the sea, at a distance of five or six miles, which *traveled toward the north-east*. On passing over Strousa, the wind from a slight air became *a hurricane*, and a thick, well-defined belt of large hailstones was left on the island. The barometer fell two inches." — (Transactions Royal Society, Edinburg, vol. ix.)

2d. Hurricanes, whirlwinds, and hailstones accompany the paroxysms of volcanos.

"1730. A great volcanic eruption at Lancerote Island, and *a storm*, which was equally new and terrifying to the inhabitants, as they had never known one in the country before." — (Lyell, Principles of Geology, vol. ii.)

"1754. In the Philippine Islands a terrible volcanic eruption destroyed the town of Taal and several villages. Darkness, hurricanes, thunder, lightning, and earthquakes, alternated in frightful succession." — (Edinburgh Philosophical Journal.)

"In 1805, 1811, 1813, and 1830, during eruptions of Etna, caravans in the deserts of Africa perished by violent whirlwinds. In 1807, while Vesuvius was in eruption, a whirlwind destroyed a caravan." — (Rev. W. B. Clarke in Tasw. Journal.)

"1815, Java. A tremendous eruption of Tombow Mountain. Between nine and ten P.M., ashes began to fall, and soon after *a violent whirlwind* took up into the air the largest trees, men, horses, cattle, etc." — (Raffles' History of Java.)

"1817, Dec. Vesuvius in eruption. In the evening *a hail storm*, accompanied with red sand." — (Journal of Science, vol. v.)

"1820, Banda. A frightful volcanic eruption, and in the evening an earthquake and a violent hurricane." — (Annales de Chimie.)

"1822, Oct. Eruption of Vesuvius. Toward its close the volcanic thunder-storm produced an exceedingly violent and abundant fall of rain." — (Humboldt, Aspects of Nature.)

"1843, Jan. Etna in eruption. Violent hurricanes at Genoa, in the Bay of Biscay, and in Great Britain.

"1843, Feb. Destructive earthquakes in the West Indies, a volcanic eruption at Guadaloupe, followed by hurricanes in the Atlantic."

"1846, June 26. Volcano of White Island, New Zealand, in eruption. Heavy squalls of wind and hail; it blew as hard as in a typhoon." — (Commodore Hayes, R.N., in Naut. Mag., 1847.)

"1847, March 20. Volcanic eruption and earthquake in Java; and on the 21st of March, and 3d of April, violent hurricanes." — (Java Courant.)

"1851, Aug. 5. A frightful eruption of the long dormant volcano of the Pelée Mountain, Martinique. Aug. 17. Hurricane at St. Thomas, etc.; earthquake at Jamaica, etc.

"1852, April 14. Earthquake at Hawaii, and on the 15th a great volcanic eruption. On the 18th *a gale of unusual violence* lasted thirty-six hours, and did great damage." — (The Polynesian, April 22, 1852.)

3d. In volcanic regions, earthquakes and hurricanes often occur almost simultaneously, but in no certain order, and without any volcanic eruption being observed.

In 1712, 1722, 1815, and 1851, earthquakes and hurricanes occurred together at Jamaica; in 1762 at Carthagena; in 1780 at Barbadoes; in 1811 at Charleston; in 1847 at Tobago; in 1837 and 1848 at Antigua; in 1819, an awful storm at Montreal, rain of a dark inky color, and a slight earthquake. People conjectured that a volcano had broken out. In 1766 the great Martinique hurricane, a *waterspout* burst on Mount Pelée and overwhelmed the place. Same night, an earthquake.

1843, Oct. 30. Manilla. — Twenty four hours' rain and two heavy earthquakes. 10 P.M., a severe hurricane.

"1852, Sept. 16. Manilla — An earthquake destroyed a great part of the city; many vessels wrecked by a great hurricane in the adjacent seas, between the 18th and 26th of September." — (Singapore Times.)

"1731, Oct. Calcutta. — Furious hurricane and violent earthquake; 300,000 lives lost."

"1618, May 26. Bombay.—Hurricane and earthquakes; 2,000 lives lost."—(Madras Lit. Tran., 1837.)

"1800. Ongole, India, and in 1815, at Ceylon, a hurricane and earthquake shocks."—(Piddington.)

"1348. Cyprus.—An earthquake and a frightful hurricane."—(Hecker.)

"1819. Bagdad.—An earthquake and *a storm*—an event quite unprecedented.

"1820, Dec. Zante.—Great earthquake and hurricane, with manifestations of a submarine eruption."—(Edinburg Phil. Journal.)

"1831, Dec. Navigator's Islands.—Hurricane and earthquakes."—(Williams' Missionary Enterprise.)

"1848, Oct., Nov. New Zealand.—Succession of earthquake shocks, and several tempests.

"1836, Oct. At Valparaiso, a destructive tempest and severe earthquakes."—(Nautical Magazine, 1848.)

When an earthquake of excessive intensity occurs, as at Lisbon, in 1755, the volcanic craters, which act as the safety-valves of the regions in which they are placed, are supposed to be sealed up; and it is a remarkable and highly-suggestive fact, that *no hurricane follows such an earthquake*. The number of instances of the concurrence of ordinary earthquakes and hurricanes might easily be increased, but the preceding suffice to show the *generality* of their coincidence, both as *to time* and place.

4th. The breaking of water-spouts on mountains sometimes accompanies hurricanes.

In 1766, during the great Martinique hurricane, before cited.

"1826, Nov. At Teneriffe, enormous and most destructive water-spouts fell on the culminating tops of the mountains, and a furious cyclone raged around the island. The same occurred in 1812 and in 1837."—(Espy and Grey's Western Australia.)

"1829. Moray.—Floods and earthquakes, preceded by water-spouts and a tremendous storm."—(Sir T. D. Lander.)

"1826, June. Hurricanes, accompanied by water-spouts and fall of avalanches, in the White Mountains."—(Silliman's American Journal, vol. xv.)

5th. The fall of an avalanche sometimes produces a hurricane.

"1819, Dec. A part (360,000,000 cubic feet) of the glacier fell from the Weisshorn (9,000 feet). At the instant, when the snow and ice struck the inferior mass of the glacier, the pastor of the village of Randa, the sacristan, and some other persons, *observed a light*. A frightful hurricane immediately succeeded." — (Edinburg Philosophical Journal, 1820.)

6th. Water-spouts occur frequently near active volcanos.

This is well known with regard to the West Indies and the Mediterranean. The following notices refer to the Malay Archipelago and the Sandwich Islands:

"Water-spouts are often seen in the seas and straits adjacent to Singapore. In Oct., 1841, I saw *six* in action, attached to one cloud. In August, 1838, one passed over the harbor and town of Singapore, dismasting one ship, sinking another, and carrying off the corner of the roof of a house, in its passage landward." — (Journal of Indian Archipelago.)

"1809. An immense water-spout broke over the harbor of Honolulu. A few years before, one broke on the north side of the island (Oahu), washed away a number of houses, and drowned several inhabitants." — (Jarves' History of Sandwich Islands.)

7th. Cyclones begin in the immediate neighborhood of active volcanos.

The Mauritius cyclones begin near Java; the West Indian, near the volcanic series of the Caribbean Islands; those of the Bay of Bengal, near the volcanic islands, on its eastern shores; the typhoons of the China Sea, near the Philippine Islands, etc.

8th. Within the tropics, cyclones move toward the west; and, in middle latitudes, cyclones and water-spouts move toward the N. E., in the northern hemisphere, and toward the S. E. in the southern hemisphere.

9th. In the northern hemisphere, cyclones rotate in a horizontal plane, in the order N. W., S. E.; and in the southern hemisphere, in the order N. E., S. W.

By applying the principles of electro-dynamics to the electricity of the atmosphere, I shall endeavor to connect and explain the preceding well-defined facts. The continuous observations of Quetelet, on the electricity of the atmosphere, from 1844 to 1849 (Literary Journal, February, 1850), show that it is always positive, and increases as the temperature diminishes. It therefore increases rapidly with the height above the earth's surface. We may, consequently, regard the upper and colder regions of the atmosphere as an immense reservoir of electric fluid enveloping the earth, which is insulated by the intermediate spherical shell formed by the lower and denser atmosphere. Now, whenever a vertical column of this atmosphere is suddenly displaced, the surrounding aqueous vapor will be immediately condensed and aggregated, and the cold rarefied air and moisture will form a vertical conductor for the descent of the electrical fluid. This descent will take place down a spiral, gyrating in the order N. W., S. E., in the northern hemisphere, since the electric current is under the same influence as that of the south pole of a magnet; and in the order N. E., S. W., in the southern hemisphere. The air exterior to the conducting cylinder will partake of the violent revolving motion, and a tornado or cyclone will be produced.

Upon the foregoing facts I shall comment in another place.

Three theories have been advanced by meteorologists of this country, two of which profess to explain all the phenomena of the weather. Professor Espy attributed the production of storms and rain to an ascending column of air, rarefied by heat, and the rarefaction increased by the latent heat of vapor given out during condensation, and an inward tendency of the air, from all directions, toward the ascending vortex, constituting the prevailing winds. Thus, Professor Espy conceived, and to some extent proved, that the wind blew inward, from all sides, toward the center of a storm, either as a circle, or having a long central line, and he conceived that it ascended in the middle, and spread out above; and that clouds, rain, hail, and snow, were formed by condensation consequent upon the expansion and cooling of the atmosphere, as it attained an increased elevation.

This ascent was not, in fact, *proved* by Professor Espy, *has not been found by others*, and *is not discoverable, according to my observations*. The theory was ingenious, founded on the theory of Dalton, that the vapor was maintained in the atmosphere by reason of a large quantity of latent heat, which was given out when condensation took place. This theory is also unsound. No such elevation of temperature is found in clouds or fogs when they form near the earth, however dense. Thus the two principal elements of Professor Espy's theory are found to be untrue, and the theory untenable. But it was sustained with great ability and research, and the distinguished theorist deserves much for the discovery and record of important facts in relation to the weather. Aside from its theoretical views, his book contains a great mass of valuable information, and will well repay the cost of purchase and perusal.

Another theory, by Mr. Bassnett, is of recent date, founded on the influence of the moon, and the supposed creation of vortices in the ether above, whose influence extends to the earth, producing storms and other phenomena. No one can peruse his book without conceding to him great ability and scientific attainment; and if his theory was true, the periods of fair and foul weather could be calculated with great mathematical certainty. But it contains inherent and insuperable objections. I will only add that all herein before contained is in direct opposition to it.

Mr. W. C. Redfield, of New York, as early as 1831, first advanced in this country the theory of gyration in storms, and investigated their lines of progress on our coast and continent. His theory is limited in its character, and does not profess, except indirectly, to explain all, or indeed any, of the other phenomena of the weather. As far as it goes, however, it is generally received in this country and

Europe, and has been adopted by Reed, Piddington, and others, who have written on the law of storms. The position of Mr. Redfield is honorable to himself and his country. Science and navigation are much indebted to him for his industry in the collection of facts. Nevertheless, his theory is not in accordance with my observation, and I deem it unsound. Although expressed disbelief of the theory has been characterized as an "attack" upon its author, I propose, with that *respect* which is due to him, but with that *freedom* and *independence* which a search for *truth* warrants, to examine it with some particularity. It is a part of the subject, and I can not avoid it.

When the theory was first announced, I adopted it as probably true; and being then engaged in a different profession, which took me much into the open air by night and day, I watched with renewed care the clouds and currents for evidence to confirm it. I discovered none; on the contrary, I found much, very much, absolutely and utterly inconsistent with its truth. The substance only of these observations will be adduced.

Mr. Redfield admits that the progression of our storms in the vicinity of New York, is from some point between S. S. W. and W. S. W., to some point between N. N. E. and E. N. E. According to my observation, except perhaps in occasional autumnal gales, they are not often, if ever, from S. of S. W., and the great majority of them, including, I believe, all N. E. storms, are between S. W. and W. S. W. Now, the card of Mr. Redfield, moving over any place from any point between S. W. and W. S. W., calls for a S. E. wind at its axis, an E. wind at its north front, and a S. wind at its south front, and does not call *for a N. E. wind on its front at all, except at the north extreme*, where it could *not continue for any considerable period.*

In relation to this, I observe, 1st. *About one-half of our N. E. storms, including some of the most severe ones, not only set in N. E., but continue in that quarter without veering at all, during the entire period that the storm cloud is over us;* usually for twenty-four hours; not unfrequently for forty-eight hours, sometimes for seventy-two or more hours. This every one can observe for himself, and it can not, of course, be reconciled with his theory.

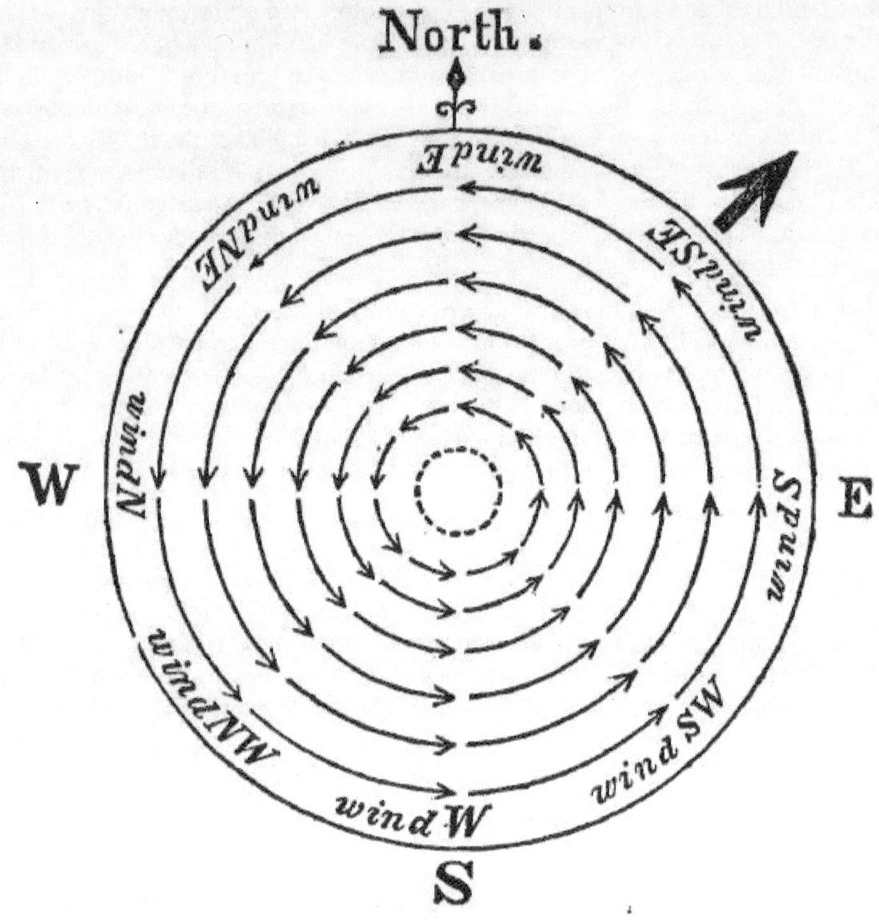

FIG. 17.

2d. N. E. storms, whether they set in from that quarter in the commencement, or veer to it afterward, when they do "change" round, more frequently veer by the S. to the S. W. in clearing off, than back through the N. into the N. W. The former, in accordance with his theory, they can not do, as the reader can see by passing the left side of the card over his place of residence on the map from S. W. to N. E.

3d. N. E. storms often pass off without hauling by S. or backing by N., and with or without a clearing off shower, the *wind shifting and coming out suddenly at S. W.* This they could not do in accordance with his theory, as slipping the card will show.

4th. From June to February it is *exceedingly uncommon* for a N. E. storm to back into the N. W. They do so more frequently from February to May, especially about the time of the vernal equinox and after; and then, because the focus of precipitation and storm intensity of the extra tropical zone of rains is S. of 42° east of the Alleghanies. His theory requires them to back by N. into N. W. *in all cases, when they set in N. E.*

5th. When they do back from the N. E. into the N. W., it rarely indeed continues to storm after the wind leaves the point of N. E. by N., and generally, if it does continue stormy, *the wind is light*, and not a gale, how violent soever the gale from the eastward may have been. Usually, by the time the wind gets N. W., it has cleared off. This, Mr. Redfield, as we shall see, evades by embracing the N. W. fair wind as a part of the same gale. According to my observation, therefore, a *very large proportion* of the *N. E. storms*, and they are a majority of the most violent ones of our climate east of the Alleghanies, do not *commence, continue*, or *veer* in accordance with his theory, but the *reverse*; and so long as this is so, I can not receive his theory as true.

6th. S. E. storms do not always, or indeed often, conform to the requirements of his card. When they set in violently at S. E., and continue so for hours without veering, the axis of the storm should be over us, and the wind should change *suddenly* to N. W. This did not occur in the storm of Sept. 3, 1821, nor does it often, if ever, occur in the summer or early gales of the autumnal months. In the later storms of autumn, and as often in those which are very gentle as any, and in the winter months when S. E. gales are rare, it does sometimes so change after the storm cloud has passed. But in the winter months, as in the storm investigated by Professor Loomis, the storms are frequently long from S. E. to N. W., and the S. E. wind blows nearly in coincidence with its long axis, for a thousand or fifteen hundred miles, till the barometric minimum is passed, and the inducing and attracting force of this part of the storm cloud is spent, and then the N. W. wind follows; sometimes blowing in under the storm cloud, turning the rain to snow; but oftener following the storm within a few hours, or the next day. The storm of Professor Loomis, when over Texas, was not probably more than four or five hundred miles in length. As it curved more, and passed north and east, it extended laterally, its center traveling with most rapidity, and when it reached the eastern coast was about fifteen hundred miles long, and not more than six hundred broad. Along the eastern part of that storm, except when by its more rapid progress the front projected much further eastward over New England than its previously existing line, the S. E. winds blew. When it bulged out, so to speak, by reason of the increased progress of the center, the wind veered to the N. E. The center of the storm passed near St. Louis and south of Quebec, as the *fall of rain*, the *bulging* of the *rapidly-moving center*, and the *line of subsequent cold*, attest. It is utterly impossible for any unbiased mind to look at the description of that storm, and attribute to it a rotary character. With all the data before him, Mr. Redfield himself has not attempted it directly.[8]

The September storm of 1821 was more violent in character than any which have since occurred. My recollection of it is as distinct as if it occurred yesterday. Peculiar circumstances, not important in this connection, fixed my attention upon the weather during that day and night. There were cirro-stratus clouds passing all day, from about S. W. to N. E., thickening toward night with fresh S. S. W. wind and flocculent scud, such as I have since seen at the setting-in of S. E. autumnal

---

[8] See Am. Jour. of Science, New Series, Vol. 18. p. 187.

gales. In the evening the wind (in the immediate neighborhood of Hartford, Ct.), veered to S. E., the cloud floated low, it became very dark, and the wind blew a most violent gale. The trees were falling about the house where I then resided, the windows were burst in, and I was up and observant. When the cloud passed off to the east, it was suddenly light, and almost calm. The western edge of the storm cloud was as perpendicular as a steep mountain side, and was enormously elevated, and very black. I have sometimes seen the western side of a summer thunder cloud, which had drawn a violent gust along beneath it, as elevated and perpendicular, but never a storm cloud. No cloud of that *depth*, or *intensity* as exhibited by its peculiar blackness, ever floated or will float so near the earth, without inducing a devastating current beneath. After it had passed the ridges east of the Connecticut valley, its top could be seen for a long and unusual period over the elevated ranges.

Now that storm was but an *intense portion* of an extensive stratus-rain cloud. Such portions frequently exist, and Mr. Redfield admits the fact. Another like portion, in the same storm, passed over Norfolk, Virginia, and the adjacent section, where the wind was N. E., and veered round by N. W. to S. W. Baltimore, and some vessels at sea, were between the two intense portions of the storm, and were not affected by either. Its northern limit was bounded by a line, drawn from some point not far north of Trenton, New Jersey, north-eastward, and north of Worcester, Massachusetts. I was about forty miles south of its northern limit, and north of its center. During that day, and the next, there was wind from S. W. to S. E., inclusive, including the gale, and *from no other quarter*. It did not at any time veer to the W. or N. W. After the passage of the storm-cloud, the wind was very light. When this intense portion of the storm passed over the valley of the Connecticut, its longest axis was from S. S. E. to N. N. W., and the *wind was S. E. the whole length of it*. In its passage from the longitude of Trenton to Boston, there was N. W. wind at one point, and but one, and that was in the iron region, at the N. W. corner of Connecticut, at the northern limit of the intense cloud, and owing, doubtless, to some local cause. The direction of the wind in that storm was in accordance with what is generally true of our storms. The wind on the front of the storm depends upon its shape. If the storm is long in proportion to its width (and no other *violent* autumnal or winter storm has been investigated, to my knowledge), the wind blows axially, or obliquely, on its front. Thus, if long from S. E. to N. W., the wind on its front will blow from the S. E. So, if the storm is long from S. W. to N. E., and has a south-eastern lateral extension, with an easterly progression, the wind will blow axially in the center, and obliquely at the edges. Instances might be multiplied, but I refer to one of recent date and striking character. All of us remember the drought of 1854. It ended in drenching rain on the 9th of September. This rain fell from a belt, half showery and half stormy in character, which had a S. E. lateral extension.

The evening of the previous day there was some lightning visible at the north, and the usual S. S. W. afternoon wind *continued fresh after nightfall*. The next day we had a brisk wind from the same quarter, and, after noon, the clouds appeared to pile up in the far north, seeming very elevated. They continued to do so, extending southerly during the afternoon, *with a high wind from S. S. W.*, the cumulus clouds moving E. N. E. At 5 P.M., gentlemen who left New York at 3 P.M., reported that a dispatch had been received from Albany, dated 1 P.M., stating that it was raining very heavily there. About 7 P.M., the belt reached us, and it rained heavily from that time till morning. Not far from 8 P.M., and during the heaviest rain, the wind shifted from the S. S. W. to N. E., and blew fresh and cold from that quarter during the night, and till the belt had passed south, and then from N. E. by N., cool, with heavy scud, during the forenoon, veering gradually to the N. N. E., and dying away. After the rain ceased, the northern edge of the belt was distinctly visible in the S. and S. E., its stratus-cloud moving E. N. E., and its scud to the westward.

The front of that storm did not pass over us. It was long and narrow. The wind blew somewhat obliquely inward, along its southern border, to the eastward, and, in like manner, to the westward, on its northern border, but from the N. E. axially along its central portions.

In the last instance, the wind changed from S. W. to N. E. This, too, is impossible, according to Mr. Redfield's theory. Similar instances, in summer, and early autumn, are not uncommon. But I shall recur to this in connection with the different *classes* of storms.

Again, the manner in which these S. E. winds co-exist with the N. E., and become the prevailing wind, toward the close of the storm, is instructive, and inconsistent with the theory of Mr. Redfield. In the West Indies, the first effect of the storm is to increase the N. E. trade; the wind then becomes baffling, but settles in the N. W. or N. N. W., *in direct opposition to the admitted progress of the storm*. At this point, or at S. W., it blows with most force. Sometimes it veers gradually, and sometimes falls calm, and comes out from the S. W., blowing violently. It ends by veering to the S. E., following gently the course of the storm. Thus, Mr. Edwards, in the third volume of his History of Jamaica, as herein before cited, *"all hurricanes begin from the north, veer back to W. N. W., W., and S. S. W., and when they get round to S. E. the foul weather breaks up."*

A short, sudden gale, resembling those of our summer thunder-showers, is sometimes met with from the S. E.; but the violent hurricanes of any considerable continuance are, in almost every case, as just stated.

Now, there is, in our latitudes, an obvious law on the subject, and it is this:—If the storm is not disproportionately long, northerly and southerly, there is a general tendency to induce and attract a surface current, in opposition to the course of the storm on its front, and especially its north front. At the same time, there is a tendency to induce a lateral current on its side, particularly the southerly side, and sometimes its south front: that the latter current is, in the first part of the storm, above the former; in the middle and latter part, it becomes the prevailing current at the surface, and the wind changes accordingly, with or without a calm—that this lateral change sometimes takes place on either side, but usually occurs on the side where the water is warmest, or there is, for other and local reasons, a *greater susceptibility in the atmosphere to inductive and attractive influence*. Thus, our N. E. storms very frequently have a southerly current also, drawn from the ocean, south of us, which forms the middle current, and, in the middle and latter part of it, becomes the prevailing one. *I have seen more than a hundred such instances, clearly and distinctly marked.* Since I have been writing this chapter, January 29th, 1855, such an instance has occurred. On Sunday, the 28th, the cirro-stratus were all day passing from the S. W. to N. E., and gradually thickening with light air from the E. N. E., in the afternoon. During the evening the wind set in *violently* from the N. E., with a deluging rain. During the night, and after a brief calm, it changed suddenly to the southward, and blew in like manner. This morning the storm was gone, and with it, six inches of hard, frozen icy snow; the trade was clear, with the exception of here and there a broken, melting piece of stratus, but scud were still running from the southward, and the wind has been from the south, veering to S. W., all day, with sunshine. As I have before remarked, this middle current is always present, in this locality, in stratus storms, when there is a heavy fall of rain or snow, although, when the latter happens, the middle current is sometimes from the northward; if it be from the southward, it turns the snow first into very large flakes, and then to rain in our part of the storm.

Doubtless, the same thing occurs every where. In the West Indies, and especially over the Leeward Islands, the middle current is most commonly from the stream of warm water which runs off to the westward into the Caribbean Sea; as the S. W. moonsoon is from the same current below the Cape de Verdes. The S. W. winds, which come from those south polar waters, in the West Indies, appear to be the most violent. But it may be on either or both sides.

The hurricane cloud of the West Indies moves confessedly N. W. in most instances, and undoubtedly it does in all. There is an immutable law that requires it. The seeming exceptions are not such; they are but instances imperfectly investigated. Now, a circular storm moving N. W. can set in N. W. only on the left front, and *can not change to S. W. on that side of the axis.* Nor can the wind blow at the axis from N. W. at all. It should be N. E. in first half, and S. W. in last half. Strange as it may seem, the axis of a West India hurricane in conformity with Mr. Redfield's theory, and a N. W. progression, has never been found, with perhaps a single exception, in any one of which I have seen a description. On the west coast of Europe, the gale is commonly from the Atlantic, either following under the storm from the S. W., or blowing in diagonally from the W. or N. W.; the N. E. wind of western Europe being a cold, dry wind, which there is reason to believe has been around the Siberian pole and is returning, as the cold northerly winds of the North Pacific have around the North American magnetic pole. "If the N. E. winds always prevailed," says Kämtz, speaking of Berlin, "even at a considerable height it would never rain." This was based on an observation of showers, and not fully reliable. But the dry and cool character of the N. E. wind of western Europe is unquestionable. The S. E. wind is also a storm wind, but owing to the character of the surface from which it is attracted, it is not as violent as the westerly winds are.

Such, too, is the general course and character of the side wind in the southern hemisphere. There gales are less frequent, the magnetic intensity is less, the counter-trades are less; it is not in "the order of Providence" that as much rain shall fall there. Nevertheless, gales occur, although rarely, if ever, with equal violence. About New Holland, where storms are pursuing a S. E. course, they have the wind N. E., corresponding to our S. E., veering from thence, *by the north*, to the westward, clearing off from S. W., with a rising barometer, as ours do from N. W.

In the Bay of Bengal, the Indian Ocean, and the Arabian Sea, there is more irregularity.

But the law of progress and lateral winds can be distinctly traced as *present* and prevailing, notwithstanding the irregularities. Our limits do not permit an analysis. In the celebrated case of the Charles Heddle, there was much evidence to show that she was driven across the front of the storm by one lateral wind, and back by another. (Diagram of Colonel Reid, p. 206.)

The waters of the Indian Ocean are hot and confined. Storms there are often composed of detached masses, move slower—sometimes not more than three or four miles an hour—and they curve over the ocean, where it is hotter than in any similar latitude. Yet, notwithstanding all peculiarities and irregularities, the law we have been considering is probably the *prevailing* law there.

No man knows better the existence of these different currents than Mr. Redfield. Doubtless it has escaped his attention that the upper of two, after the passage of a considerable proportion of the storm, becomes the lower, and causes a seeming change of the same wind.

In a series of elaborate articles, substantially reviewing the whole subject, published in the American Journal of Science, for 1846, he says:

> "In nearly all great storms which are accompanied with rain, there appear two distinct classes of clouds, one of which, comprising the storm scuds in the active portion of the gale, has already been noticed. Above this is an extended stratum of stratus cloud, which is found moving with the general or local current of the lower atmosphere which overlies the storm. It covers not only the area of rain, but often extends greatly beyond this limit, over a part of the dry portion of the storm, partly in a broken or detached state. This stratus cloud is often concealed from view by the nimbus, and scud clouds in the rainy portion of the storm, but by careful observations, may be sufficiently noticed to determine the general uniformity of its specific course, and, approximately, its general elevation.

> "The more usual course of this extended cloud stratum, in the United States, is from some point in the horizon between S. S. W. and W. S. W. Its course and velocity do not appear influenced in any perceptible degree by the activity or direction of the storm-wind which prevails beneath it. On the posterior or dry side of the gale, it often disappears before the arrival of the newly condensed cumuli and cumulo-stratus which not unfrequently float in the colder winds, on this side of the gale."

> "The general height of the great stratus cloud which covers a storm, in those parts of the United States which are near the At-

lantic, can not differ greatly from one mile; and perhaps is oftener below than above this elevation. This estimate, which is founded on much observation and comparison, appears to comprise, at the least, the limit or thickness of the proper storm-wind, which constitutes the revolving gale.

"It is not supposed, however, that this disk-like stratum of revolving wind is of equal height or thickness throughout its extent, nor that it always reaches near to the main canopy of stratus cloud. It is probably higher in the more central portions of the gale than near its borders, in the low latitudes, than in the higher, and may thin out entirely at the extremes, except in those directions where it coincides with an ordinary current. Moreover, in large portions of its area, there may be, and often is, more than one storm-wind overlying another, and severally pertaining to contiguous storms. In the present case, we see, from the observations of Professor Snell and Mr. Herrick, at Amherst, Massachusetts, and at Hamden, Maine (115 and 135 b.), that the true storm wind, at those places, was super-imposed on another wind; and various facts and observations may be adduced to show that brisk winds, of great horizontal extent, are often limited, vertically to a very thin sheet or stratum."

Much of the foregoing is graphically described, and unquestionably true. But it may well be asked how he, or others, distinguish which of two or more currents (for there are frequently three, and sometimes four visible), are the true currents of the storm, and which interlopers from another storm? Is the true one always the upper one, and why? If the upper one, why is the interloper at the surface noted and quoted to prove what a storm is? How does he know what proportions of the winds he has recorded to show the revolving motion of gales, were the true storm winds of the particular storm? or, that every one of them was not an interloping wind on which the true storm wind was superimposed?

These inquiries are pertinent, for obviously, unless some rule for distinguishing between the currents is given, and there be evidence of direct observation to show that the surface wind, whose direction is noted, is the true wind of the storm, and that the *latter* is not *superimposed*, no reliance can be placed upon logs, or newspaper accounts, or registers. There is another element besides direction, viz.: superimposition, a determination of which *is* essential to *truth*. It will be difficult for Mr. Redfield to say that a determination of that element has been made, with certainty, in a single storm he has investigated; and in relation to the convergence of storms, and blending, and superimposition of their winds, I think he is mistaken.

Mr. Redfield is right in saying (American Journal of Science, vol. ii., new series, p. 321) that "too much reliance may be placed upon mere observations of the surface winds in meteorological inquiries," and yet *they* only have thus far been regarded, and he has proved gyration in no other way. I have frequently, with a vane in sight, asked intelligent men how the wind was, and been amused and instructed by their inability to state it correctly. Mr. Redfield, in his inquiries, often found two reports of the weather at the *same time*, from the *same place*, materially different; and I have known, from my own observation, newspapers and meteorological registers to be several points out of the way; and this, because the vanes are influenced by local elevations, and change several points, and very often; because few know the exact points of the compass in their own localities, and because entire accuracy has not been deemed essential. For these reasons, newspaper and telegraphic reports are not always reliable; and therefore, and because, also, storm-winds are easterly and fair winds westerly, and the former veer from east around to west, on one or both sides in many cases, there are few storms which can not be represented as whirlwinds, by a proper *selection* of *reports*, a corresponding *location* of the *center*, and an *extension* of the lines of supposed gyration, so as to include the *preceding* winds, the actual winds of the storm, and the *lateral*, and *succeeding* fair weather ones.

But, again, Mr. Redfield is right in saying there is, in such cases, "an extended stratum of stratus cloud," and it is always present. But why does he say this *covers the storm*? Is it distinct from it, and if so, what is it doing there? What power placed it there, and for what purpose? Has this extended stratum of cloud, which forms the canopy of a vast chamber—five hundred to one thousand miles in diameter, and less than two miles in vertical depth, while the earth forms the floor—any agency in producing the whirl that is supposed to be going on within it, and if so, what? Has the earth any agency, and if so, what? If neither the ceiling nor floor of the chamber have any agency in producing it, what does? Are we to consider the *storm-scud* as possessing the power, and as waltzing around the aerial chamber, carrying the air with them in a hurricane-dance of devastation? *What, in short, is the power, and how is it exerted?*

To these questions, Mr. Redfield's essays furnish no comprehensive answer. There is an intimation that the cause of storms will be, at some future day, developed. One attempt, and but one, has thus far been made, and that I quote entire:

> "We have seen that the two Cuba storms, as well as the Mexican northers, have appeared to come from the contiguous border of the Pacific Ocean.
>
> "Now, are there any peculiarities in the winds and aerial currents of those regions, which may serve to induce or support a leftwise rotation in extensive portions of the lower atmosphere, while moving on, or near the earth's surface? I apprehend there are such peculiarities, which have an extensive, constant, and powerful influence. First, we find on the eastern portion of the Pacific, from upper California to near the Bay of Panama, an almost constant prevalence of north-westerly winds at the earth's surface. Next, we have an equally constant wind from the southern and south-western quarter, which, having swept the western coast of South America, *extends across the equator to the vicinity of Panama*, thus meeting, and commonly over-sliding the above-mentioned westerly winds, and tending to a deflection or rotation of the same, from right to left. As this influence may thus become

extended to the Caribbean or Honduras Sea, we have, next, the upper or S. E. trade of this sea, which is here frequently a surface-wind, and must tend to aid and quicken the gyrative movement, ascribed to the two previous winds; and lastly we have the N. E. or lower trade, from the tropic, which, coinciding with the northern front of the gyration, serves still further to promote the revolving movement which may thus result from the partial coalescence of these great winds of Central America, and the contiguous seas.

"Thus, while a great storm is, in part, on the Pacific Ocean, its N. E. wind may be felt in great force on that side of the continent, through the great gorges or depressions near the bays of Papagayo or Tehuantepec, as noticed by Humboldt, Captain Basil Hall, and others, the elevations which there separate the two seas being but inconsiderable; and, when the gyration is once perfected, the whole mass will gradually assume the movement of the predominant current, which is generally the higher one, and will move off with it, integrally, as we see in the cases of the vortices, which are successively found in particular portions of a stream, where subject to disturbing influences."

The analogy between this and the theory of Professor Dove, cited above, and prior, in point of time, is obvious. They are substantially alike in principle, with different locations. They differ also in this, Professor Dove appears to think something more than over-sliding necessary, and assigns the duty of crowding the upper current down in to the lower, to make an *encounter*, to a lateral overflow from Africa. Mr. Redfield seems to think there may be a tendency to deflection when they "over-slide" each other. They are both closet hypotheses, the poetry of meteorology, with something more than poetical license as to facts.

In the first place, *no such concurring winds exist in the same locality at the same time*. When the inter-tropical belt of rains is over Central America and Southern Mexico, a S. W. monsoon blows in under it, but it usurps the place of all other surface winds; and, when the belt is absent, that portion of the eastern Pacific is most remarkably calm, or is covered by the N. E. trades. Secondly, the *trade-winds every where pursue their appointed course without "tendency to deflection" by the meeting, or "over-sliding," or "breaking in," or "encounter,"* of other winds. The great laws of circulation do not admit of any such *confusion*. And, lastly, *no storm ever came over the eastern United States from that quarter*. The unchangeable laws of atmospheric circulation forbid it. Recent observations also have shown that the storms on the west coast of Central America, and the eastern Pacific, pursue a N. W. course, precisely as in the West Indies, and every where over the surface-trades of the northern hemisphere. Indeed *Mr. Redfield himself has recently investigated several of them, and admits their course to be north-westerly.* (See American Journal of Science, new series, vol. xviii. p. 181.)

But, suppose the co-existence of the winds and the course of the storms admitted as claimed, let us seek for clearer views. What do these gentlemen mean?

Do they intend to have us believe the air has inherent moving power, and that the "tendency" of which they speak is an attribute of the winds, and that when they thus meet, and "come into each other," "encounter," or "over-slide," and become acquainted, they wheel into a waltz, and move off northward, "integrally," with unceasing circular movement, even until they arrive at the Arctic circle? Or is it a mere mechanical effect of meeting, "coming into each other," or "over-sliding?" If the latter, why a tendency to rotation from right to left? The trade-winds, at least, are *continuous, unbroken sheets*, and not disconnected portions which meet and blow past each other, and there is no warrant for placing them *side and side*, and attributing to them any such mechanical effect, and as little respecting the other winds. Outside of the fanciful hypothesis, there are no facts to show such a tendency one way rather than the other; and, in accordance with the known facts regarding stratification of the currents of air, no such "tendency" can exist.

But what *power* impels the winds, which thus meet at these points? If they be impelled, is it consistent with the action of this power that the *winds* it has *created* and *controls*, should thus assume an *opposite "tendency,"* and whirl away to the north-eastward, regardless of the power that originated and controls them? What must this *"tendency"* be, which thus *occasionally* not only diverts the winds from the *usually regular course* given them by their originating power, but increases their action, from gentle, ordinary winds, to hurricanes? Nay, which gives them a new, resistless gyratory and electric energy, increasing as the new, independent, supposed cyclonic organization moves off, *"integrally,"* away from "the home of its many fathers," on a devastating journey towards the north pole?

And, further, if all this were true as to the West Indies and Central America, what is to be said of the billions of other storms, originating on a thousand other portions of the earth's surface, and how are they to be accounted for, inasmuch as such other "meetings," "coming into each other," and "over-sliding," and "tendency to deflection," is not assumed to exist?

These questions cannot be satisfactorily answered. The distinguished theorists are mistaken. The stratus-cloud does not over-lie or cover the storm. It is the storm. The winds beneath, whether surface or superimposed, are but its incidents, due to its static induction and attraction. Their *direction* depends on the shape of the storm cloud, and its course of progression, and the susceptibility of the surface atmosphere in this direction or that, to its inductive and attractive influence. Their *force* to its depth, its contiguity to the earth, and the intensity of its action; and the scud, are but patches of condensation, occasioned by the same inductive action which affects and attracts the surface current in which they form.

Another objection to Mr. Redfield's theory of gyration is based upon the fact that in order to constitute his *storm*, to get the *gyration*, he has to include, at least, an equal amount, generally a great deal more, of *fair weather*. The N. W. wind, the "posterior, or dry side of the gale," as he calls it (in the foregoing extract), is a *fair weather wind*. It is *necessary*, however, to complete the supposed *circle*, and it is *pressed into the service*. The practical answer given to the question, *"what are storms?"* is, they are cyclones, part storm, so called, and *part fair weather*; that is, the stratus-cloud, the scud, the easterly wind, and rain or snow of day before yesterday, were the *wet side*, or front part of the storm, and the sunshine, clear sky, and N. W. wind of yesterday, to-day, and, perhaps, to-morrow, are the posterior or dry side. When a storm clears off from the N. W. it is not *over*, it is, perhaps, *just begun*; and, inasmuch as it storms again, very soon after the wind changes back from the N. W. to the southward, in winter, our weather then is pretty much all *storms*.

The statement of this claim seems so absurd that it may appear like injustice to make it. But gyration can not be made out without it, and it is evident in the extract quoted above; in the claim that the winter northers of the Mexican Gulf are parts of passing storms; and clearly and unequivocally advanced as a distinct proposition, as follows:

"1. The body of the gale usually comprises an area of rain or foul weather, together with another, and, perhaps equal, or greater, area of fair or bright weather." (Am. Jour. of Science, vol. xlii. p. 114.)

Now, in the first place, we must distinguish between a storm and fair weather, before we can tell what the former is, and it is difficult to assent to a theory which explains what a S. E. storm of *twelve hours'* continuance is, by including *two or three days of succeeding N. W. fair weather wind*, as a part of it. There is no proportionate relation as to *time*, nor any relation as to *qualities*, or the attending conditions of the atmosphere, nor any conceivable *connection*, except the hypothetical one of *gyration*, between the two winds.

And, in the second place, it is true, and Mr. Redfield is well aware of the fact, that winds often blow for many days from the N. E., S. W., or N. W., without any preceding or succeeding winds to which they have any discoverable relation. If, therefore, truth would justify Mr. Redfield in including the fair weather wind, a difficulty would remain which his theory does not cover or explain.

No American, except Mr. Redfield, has been able to discover satisfactory evidence of the gyration of storms, by actual careful observation, or a careful unbiased collation of the observation of others. Professor Coffin is reported to have read to the Scientific Association, at their Buffalo meeting, a paper, confirmatory, in part, but I have not been able to see it. The tracks of tornados have been searched as with candles. When they have been narrow, from forty to eighty rods, their action has been substantially similar, and, although, as we have herein before stated, some irregularities have been found which were consistent with gyration—for irregularities attend the violent action of all forces, and particularly the motion of electricity through the atmosphere, as every one who has seen the zig-zag course of a flash of lightning knows—yet the evidence of two lateral inward currents, or lines of force, has predominated over all others. In all cases, where the path is narrow, those lateral currents are the actors; they constitute the tornado; their *irregularities* of action produce the exceptions; but the exceptions are neither numerous nor uniform, and do not prove either the theory of Mr. Espy or that of Mr. Redfield. The action is not that of moving air, merely, but of a power exceeding in force that of powder, which nothing but electricity or magnetism can exert. As the path widens, the wind becomes more like the straight-line gust which follows beneath the ordinary severe thunder-showers. His theory finds no substantial confirmation or support in the path of the tornado.

Several storms were investigated by Professor Espy, some of them the same which Mr. Redfield had attempted to show were of a rotary character; one or two by the Franklin Institute of Philadelphia; one by Professor Loomis, already alluded to; and recently, two by Lieutenant Porter, from logs returned to the National Observatory. None of these investigations confirm the theory of Mr. Redfield. Indeed, Mr. Redfield himself has found it necessary to resort to suppositions of *modifying causes* to explain the evident inconsistencies. It is assumed that the axis, or center, oscillates, and describes a series of circles; and thus, one class of difficulties is avoided. Again, it is assumed that simultaneous storms converge and blend upon the same field, and another class of difficulties are surmounted. And, again, inasmuch as it is notorious that violent gales are rarely if ever felt with equal violence around the area of a circle, but from one or two points only, it is assumed, that the storm winds ascend, superimpose, and descend again, when they return to the place of their first violent action, etc. The *simple truth* requires no such resort to *modifying hypothesis.*

Still, another objection is, that the changes in the barometer, which occur before, during, and after storms, do not sustain the claims of Mr. Redfield or the requirements of his theory.

The barometer sometimes rises before storms. It generally commences falling about the time, or soon after the storm sets in, continues to fall during its progress, and rises again, sooner or later, afterward. This is the general rule.

On this subject Mr. Redfield's claim is this:

> "EFFECT OF THE GALE'S ROTATION ON THE BAROMETER.—The extraordinary fall of the mercury in the barometer, which takes place in gales or tempests, has attracted attention since the earliest use of this instrument by meteorologists. But I am not aware that the principal cause of this depression had ever been pointed out, previously to my first publication in this journal, in April, 1831, when I took the occasion to notice this result as being obviously due to the *centrifugal force* of the revolving motion found in the body of the storm.

> "Since that period, inquiries have been continued by meteorologists in regard to the periodical and other fluctuations of the barometer, and the relations of these fluctuations to temperature and aqueous vapor. But these incidental causes of variation, in the atmospheric pressure, prove to be of minor influence, and we are left to the sufficient and only satisfactory solution of this marked phenomenon which is found in the centrifugal force of rotation."

The average pressure of the atmosphere, at the surface of the ocean, or in the interior of the country, allowing for elevation, is about equal to the weight of a column of quicksilver, thirty inches in height; hence the barometer is said to stand at about thirty inches at the level of the sea.

This is sufficiently accurate for the northern hemisphere, north of the N. E. trades; but the average is somewhat lower in the trades and in the southern hemisphere. Thus, the average of sixteen months, during which the Grinnell expedition was absent, was 30.⁰⁸⁄₁₀₀.

From a large number of logs examined by Lieutenant Maury, the mean elevation in the N. E. trades of the Atlantic was 29.⁹⁷⁄₁₀₀; the S. E. trades of the Atlantic, 29.⁹³⁄₁₀₀; off Cape Horn, 29.²³⁄₁₀₀; S. E. trades of the Pacific, 30.⁰⁵⁄₁₀₀; N. E. trades of the Pacific, 29.⁹⁶⁄₁₀₀. The height of the barometer off Cape Horn is not a fair index of the general elevation of the southern hemisphere, inasmuch as it stands lower there than at the coast of Patagonia and Chili, or at most, if not all, other stations in that hemisphere.

As the barometer is constantly oscillating up and down (irrespective of its diurnal oscillation), it has no known fair weather standard. The point of 30 inches is taken only as it is a mean. I have known it to commence storming when the barometer was at 30.70, and not to fall before it cleared off, below 30.30. And I have known it to be below 30 for several days consecutively, with fair weather. In our climate there is no reliable fair weather standard for the barometer. It falls below 30 without storming; it rises far above, and storms without falling below. No reliance can be placed upon its elevation, except by comparison; but of that hereafter.

The general rule, nevertheless, is, that it falls more or less during storms, whatever its height, and rises sooner or later, more or less, after they clear off.

The difference between its highest and lowest points is called its range. The greatest range observed, and recorded, is about 3 inches—from about 28 to 31—but this range is rare. The range, in the trade-wind region, is comparatively small; in this country it is greater than in Europe; and, generally, the range will be found greatest where the volume of counter-trade, and magnetic intensity, and the corresponding amount of precipitation, and extremes of heat and cold are greatest. One of the greatest ranges during one storm, or two successive portions of a storm, in this country, which I have seen recorded, occurred at Boston, in February, 1842. It was as follows—counting the hours as 24, and from midnight:

> Feb. 15..10h..30.36.
>   "   16..13h..28.47 fall of 1.89 in 27 hours.
>   "   17..19h..30.39 rise of 1.92 in 30 hours.
>   "   18.. 2h..30.39 stationary 5 hours.
>   "   19.. 2h..29.46 fall of 0.93 in 24 hours.
>   "   20.. 2h..30.43 rise of 0.97 in 24 hours.
>   Amount of oscillation, 5.71 in 4 days, 11 hours.

These ranges were owing to the alternation of S. E. storms, and N. W. winds.

Taking the first range as a basis, and allowing the height of the atmosphere to be 1,100 feet for the first inch, we have nearly 2,000 feet displaced during one day, if we look for the displacement near the earth, or some 30 or 35 miles, if we soar aloft in the upper regions to look for the *lateral overflow* of Professor Dove, and about the same quantity restored the next. This brings us to the inquiry, how was it done? It is perfectly idle to talk about *difference* of *temperature* or *tension* of *vapor*, the *ascent* of warm air, or *descent* of cold in a case like this; or to say that they were occasioned by a lateral overflow of some thirty miles of its upper portion, first this way and then that, in such a brief space of time. The change is equal to nearly $\frac{1}{15}$ of the weight of the whole atmosphere, and the cause, whatever it was, existed within two or three miles of the earth. Mr. Redfield's explanation I give in his own words, at length:

"One of the most important deductions which may be drawn from the facts and explications which are now submitted, is an explanation of the causes which produce the fall of the barometer on the approach of a storm. This effect we ascribe to the centrifugal tendency or action which pertains to all revolving or rotary movements, and which must operate with great energy and effect upon so extensive a mass of atmosphere as that which constitutes a storm. Let a cylindrical vessel, of any considerable magnitude, be partially filled with water, and let the rotative motion be communicated to the fluid, by passing a rod repeatedly through its mass, in a circular course. In conducting this experiment, we shall find that the surface of the fluid immediately becomes depressed by the centrifugal action, except on its exterior portions, where, owing merely to the resistance which is opposed by the sides of the vessel, it will rise above its natural level, the fluid exhibiting the character of a miniature vortex or whirlpool. Let this experiment be carefully repeated, by passing the propelling rod around the exterior of the fluid mass, in continued contact with the sides of the vessel, thus producing the whole rotative impulse, by an external force, analogous to that which we suppose to influence the gyration of storms and hurricanes, and we shall still find a corresponding result, beautifully modified, however, by the quiescent properties of the fluid; for, instead of the deep and rapid vortex before exhibited, we shall have a concave depression of the surface, of great regularity: and, by the aid of a few suspended particles, may discover the increased degree of rotation, which becomes gradually imparted to the more central portions of the revolving fluid. The last-mentioned result obviates the objection, which, at the first view, might, perhaps, be considered as opposed to our main conclusion, grounded on the supposed equability of rotation, in both the interior and exterior portions of the revolving body, like that which pertains to a wheel, or other solid. It is most obvious, however, that all fluid masses are, in their gyrations, subject to a different law, as is exemplified in the foregoing experiment; and this difference, or departure from the law of solids, is doubtless greater in aëriform fluids than in those of a denser character.

"The whole experiment serves to demonstrate that such an active gyration as we have ascribed to storms, and have proved, as we deem, to appertain to some, at least, of the more violent class; must necessarily expand and spread out, *by its centrifugal action, the stratum of atmosphere subject to its influence, and which must, consequently, become flattened or depressed by this lateral movement, particularly toward the vortex or center of the storm;* lessening thereby the weight of the incumbent fluid, and producing a consequent fall of the mercury in the barometrical tube. This effect must increase, till the gravity of the circumjacent atmosphere, superadded to that of the storm itself, shall, by its counteracting effect, have produced an equilibrium in the two forces. Should there be no overlaying current in the higher regions, moving in a direction different from that which contains the storm, the rotative effect may, perhaps, be extended into the region of perpetual congelation, till the medium becomes too rare to receive its influence. But whatever may be the limit of this gyration, its effect must be to *depress* the *cold stratum* of the upper atmosphere, particularly toward the more central portions of the storm; and, by thus bringing it in contact with the humid stratum of the surface, to produce a permanent and continuous stratum of clouds, together with a copious supply of rain, or a deposition of congelated vapor, according to the state of the temperature prevailing in the lower region."

The italics in the foregoing extract are mine; and, in relation to it, I observe:

1st. There is no cylindrical vessel around storms, and *air will not thus resist air.* Confessedly, such resistance is necessary. Let any one watch his cigar smoke, and see how readily it moves on, with little momentum. Let any one try the experiment of creating a whirl in the *open air*, or in a room, or box of paper, or other material, which can be suddenly removed, with air colored by smoke. I am exceedingly mistaken if he does not find the presence of a "cylindrical vessel," absolutely essential to prevent the instantaneous tangential escape of the air.

2d. Turn back to page 3 and look at the fall of the barometer in the polar regions (recorded in the extract from Dr. Kane), with *scarcely any wind*, and *as little variation* in its *direction*, and see how utterly Mr. Redfield's theory fails to account for the phenomena.

3d. If I understand Mr. Redfield correctly, he has abandoned the claim as originally made, that the wind moves in circles, expanding, and *spreading out* by a *"lateral movement,"* and now asserts that it blows spirally inward, and elevates the air in the center. I quote:

"VORTICAL INCLINATION OF THE STORM WIND.—By this is meant some degree of involution from a true circular course. In the New England storm above referred to, this convergence of the surface-winds appeared equal to an average of about 6° from a circle. In the present case, such indication seems more or less apparent in the arrows on the storm figures of the several charts, where the concentrical circle afford us means for a just comparison of the general course of wind which is approximately shown by the several observations.

"Perhaps we may estimate the average of the vorticose convergence, as observed in the entire storm for three successive days, at from 5° to 10°—out of the 90° which would be requisite for a congeries of *centripetal* or center-blowing winds. This rough estimate of the degree of involution is founded only on a bird's-eye view of the plotted observations. But, however estimated, this involution seems to afford a measure of the air and vapor which finds its way to a *higher elevation* by means of the vortical movement in the body of the storm."

If the elevation of the air at the borders of the storm, and depression in the middle, resulted from the outward tendency and "lateral movement" of the revolving air, and from the *centrifugal force*, as in the experiment with the water in a cylindrical vessel, as stated in the first paragraph quoted, an *involution* of from 5° to 10° from the action of a *centripetal force*, must carry the air *inward*, and the *barometer should stand highest in the middle of the storm*. The change is fatal to his theory. The two are diametrically opposite in character and effect. In one, the superior strata would be brought down in the center by the *lateral pressure outward*; in the other, they would be elevated by the *involution*, which "affords a measure of the air and vapor which finds its way to a higher elevation," etc. It is perfectly obvious Mr. Redfield has refuted his own hypothesis.

In doing this, he is met by the other difficulty alluded to, which he does not attempt to explain. This gathering of the air inward, spirally, by a centripetal force, if it took place, not only would not depress, but *must elevate the barometer in the center, above that of the adjoining atmosphere.*

When he first attributed the depression of the barometer to a lateral movement and centrifugal force, he supposed the superior strata descended into the depression, and their frigidity occasioned the condensation, and cloud, and rain. How he now proposes to account for the formation of cloud and rain during storms, while the warm air of the inferior stratum finds its way to a higher elevation in the center of the storm, he does not inform us, and we must wait his time.

"I have," he says, "long held the proper inquiry to be, *what are storms?* and not, *how are storms produced?* as has been well expressed by another. It is only when the former of these inquiries has been solved that we can enter advantageously upon the latter."

The former does not seem to be yet solved, or the solution of the latter commenced. Mr. Redfield tells us (page 259, and onward), that there is an extended stratum of stratus-cloud, which overlies the storm, and that it does not differ greatly from one mile in height. We are not told how the air, which finds its way to a higher elevation during several days continuance of such a storm, *gets through the stratum.* If he is right it *must* do so, and it would not answer to *suppose* a very small opening or gentle current through it, to carry off all the air which works inward in a hurricane, during several days continuance. But he does not seem to recognize either the necessity or existence of any *vent* at all; nor is there any; and this fact is open to the observation of every school-boy in the country; and it is equally open to his observation that *when and where the barometer is most depressed, the stratus storm-cloud is nearest the earth.* Colonel Reid has much to say about the *"storm's eye,"* or "treacherous center" of a storm. A careful analysis of the instances where the "storm's eye" is noticed will show that the term is applied, in the northern hemisphere, to that lighting up in the W. or N. W., which is the commencement of the clearing-off process, and attended with a shift of wind to the fair-weather quarter: *i. e.,* to W. or N. W. Just such an "eye" as is seen when the last of the storm cloud has passed so far to the east as to admit the rays of the sun under the western or north-western edge of it. The same kind of "storm's eye" is described in the southern hemisphere, except that the wind shifts to S. W. instead of N. W., that being the clearing-off wind there. No instance of a *"storm's eye"* in the center of the extended stratum of stratus-cloud, which overlies the storm, can be found recorded, to my knowledge; and it is obvious that Colonel Reid adopts the view of Mr. Redfield, that the westerly and N. W. *fair weather* winds are a part of the storm. So long as these gentlemen hold to that opinion they will never solve the question, *"what are storms?"* or reach the other, *"how are storms produced?"*

Notwithstanding, Mr. Redfield asserts, or adopts the assertion, that the inquiry should be, "What are storms?" not "How are storms produced?" that inquiry should be a *rational* one, and should not violate all analogy, or call for an explanation which science can not *rationally* furnish. Mr. Redfield does not seem to have formed any just conception of the *immeasurable power* of a hurricane, *five hundred miles in diameter*; or of the nature of that *rod* which the *Almighty must insert in it, to whirl it with such violent and long-continued force*; nor any just conception of the tendency of the whirling mass, in the absence of his "cylindrical vessel," to fly off, tangentially, into the surrounding air; or of the nature or power of the centripetal force necessary to hold the gyratory mass in its current, and gather it in involute spirals toward a center. Nor has any other man who has witnessed, or read of mountain-tossed waves; of the largest ships blown down and engulfed; of towns submerged, and vessels carried far inland, and left in cultivated fields, by the subsidence of the sea; of sturdy forests and strongly-built edifices prostrated; or listened to the howling of the tempest, and felt his own house rock beneath him, been able to conceive of any known form of calorific or mechanical, or other power, acting from a comparatively small center, which could hold such an immense irresistable mass of whirling air in a circle, and *gather it* in toward the center in gradually contracting spirals. I confess that, to my mind, it seems little less than a mockery of our intelligence for Mr. Redfield, or Professor Dove, or any other man, how distinguished soever he may be, to tell us that all this is the result of a "tendency to left-wise rotation" of ordinary winds, "coming into each other," or "over-sliding," or "meeting," or "encountering," on this "front," or that, down in Central America, or in the West Indies, or the monsoon region; or to talk of "lateral overflows" from mere gravity; of the ascent of warm air, or the descent of cold strata; of the *resistance of adjacent passive air*, or other mere *atmospheric resistances* in connection with such *awful manifestations of power*. Their explanations of these phenomena are not rational, nor can they be believed by any rational man, who will bestow upon them half an hour of *comprehensive, unbiased reflection*.

Waiving many minor points of great force, for this notice of Mr. Redfield's theory is already too much extended for my limits, I am constrained to take issue with him on the fact, and to assert, unhesitatingly, that in a *majority of instances no such barometric curve exists*.

Doubtless the depression beneath the storm is found, and exterior lateral elevations may also be had by *extending the line into the usual fair weather elevation on each side*, as Mr. Redfield is obliged to do, to get his supposed circle of winds at all. Doubtless, too, the seamen sailing out of a storm, on either *side*, and approaching fair weather, will have a rising barometer. But from *front to rear, on the line of progression*, in tropical storms, the curve does not exist on shore, in this latitude, oftener than in two, or possibly three, cases in ten; and then only upon a single state of facts—that is, when there is an interposition of N. W. wind; and this, at some seasons, rarely occurs. An elevation usually occurs before the storm, on its front, if it present an extensive easterly front, as one of these classes does, and a *depression is left* after it has passed off, unless a considerable body of N. W. wind interposes, as heretofore stated. But when there is not such interposition of N. W. wind (for W., W. N. W., or even N. W. by W. will not suffice), there is not an immediate rise of the barometer corresponding in rapidity and extent with the fall, and frequently none during the first twenty-four hours of bright, fair weather. Let the reader, if he has access to a barometer, note this fact, for it is obvious and conclusive.

Finally, there are other atmospheric conditions to which the barometric changes are obviously due:

1st. The counter-trade is of a different *volume*, at different times, over the same locality, and hence a difference in the normal elevations of the barometer.

2d. It is at a different *elevation*, at different times, over the same locality. It was so found by the investigations of the Kew Observatory Committee referred to; has been so found by other aeronauts, and may readily be seen by a careful, practiced observer.

It is highest, with a high barometer, in serene weather, when a storm is not at hand; and can sometimes be plainly seen to ascend when a considerable volume of N. W. wind is blowing in beneath, and elevating, simultaneously, the trade and the barometer.

Opportunities occur every year, when the northern edge of the dissolving stratus-cloud is attenuated, and the storm is clearing off in the N. W., with wind from that quarter, and a rising barometer, when its gradual elevation may be observed to correspond with the *volume* of that wind.

3d. During storms, with a low barometer, the *trade* and the *clouds run low*. This, too, is clearly observable, especially when the stratus-cloud passes off abruptly, very soon after the rain ceases. In such cases the barometer will remain depressed for a considerable time, unless another storm supervenes speedily, or the wind sets in from the N. W.

4th. The *trade, in a stormy state, moves faster* than when in a normal condition. This is observable during the partial breaks which frequently occur in storms, and at other times. It is also inferable from the more rapid progress of the more intense center, and other intense portions of storms, and the consequent greater depression of the barometer, under such centers or intense portions. (See the storm of Professor Loomis.) It is obvious, also, from the greater rapidity of progress attending the more intense and violent storms which all investigations discloses.

These simple facts explain all the phenomena:

1st. The trade stratum is a continuous unbroken sheet, and its descent must displace a portion of the surface atmosphere. A portion of it is impelled forward, aiding in the precedent elevation of the barometer, and a portion is attracted backward, into the space from which a like portion had been previously attracted by the passing storm cloud, forming the easterly wind.

2d. The increased progress of the stormy portion of the counter-trade occasions an accumulation in front of the storm, and an elevation of the barometer, and tends also to increase the *depression* under the spot from which it moves. The latter is, to some extent, counteracted by the thin sheets of surface wind which are drawn in under the stratus from the sides. That which is drawn from the front in successive portions, fills the space from which like portions had been drawn to the westward, and left behind in a passive state by the passing storm. Thus, the surface atmosphere of New England may pass under the entire width of a storm, as a gale; moving now in puffs with great violence, as it passes beneath irregular and intense portions of the cloud, and now moderately; and be left, in a passive state, in Kentucky, occupying the space from which the atmosphere had been previously drawn by the same storm, *in like manner*, on to northern Texas.

3d. The nearer the stratus-cloud to the earth, the greater the displacement of surface atmosphere, the lower the barometer, and, ordinarily, the more violent the wind. First, because the same intensity, which, by attraction, brings the trade near the earth, acts with greater force upon the surface atmosphere; and, secondly, the storm winds, which are often most rapid beneath the clouds and above the earth, are likely to be felt with more violence at its surface, where the stratus cloud runs low, especially at sea.

I desire to commend all these facts, in relation to the theory of Mr. Redfield, to the careful attention and observation of those who, although believers in the theory, are not wedded to it; and who have a sincere desire to understand the phenomena which are continually, and thus far, *mysteriously*, occurring within two or three miles of us, while our knowledge of the distant worlds around us—the science of astronomy—seems almost perfect.

I will return to a further and a careful consideration of the nature of the reciprocal action between the earth and the counter-trade, and the facts bearing upon the question, in another chapter. It is obvious that received theories can not aid us materially in the inquiry.

CHAPTER X.

Further inquiry in relation to the reciprocal action between the earth
and the counter-trade—Terrestrial magnetism, and what we know
of it—Its elements, and their variations— Their connection with
the variations of atmospheric condition—Magnetism acts through
its connection with electricity—Character of the latter and its variations—Their connection with atmospheric conditions—Electricity as well as magnetism in excess over this country—Effects of it
upon our climate—Closer consideration of the atmospheric phenomena—Their diurnal changes and connections compared with
those of magnetism and electricity— Grouping of all the diurnal
variations—Particular and separate examination of them—Classification of storms— Examination in detail of the several classes
and the primary influence of the earth or counter-trade in relation
to each

We are yet ignorant of the true nature of magnetism. We trace its lines, as in
the diagrams, upon and around the magnet; but we can only do this with soft iron,
or other substance, in which magnetic action may be induced. We know that these
lines are currents, or lines of force, for that force produces sensible effects, and
we measure it by the movements of the needle. We know that these lines may be
deflected by other magnetic bodies, and concentrated upon them. We know that the
earth, and the smallest magnets, exhibit properties in common. The poles of the
magnet are some distance from its extreme ends—so are those of the earth. The
intensity increases, from the center, or near it, to the poles of the magnet, as shown
by its attraction; and the same increase of magnetic intensity, from the magnetic
equator to the magnetic poles, or near them, is traced upon the earth.

We know that there are two lines, or rather *areas*, of greater intensity upon the
globe. One extending from the American magnetic pole, south-eastwardly, to a
corresponding pole in the southern hemisphere; and another, the Asiatic, extending from the Siberian pole to a corresponding southern one, in like manner. We
know that, from those lines or areas, the intensity, east and west, on the same parallel of latitude, decreases each way, to about midway between them. Thus, calling
the intensity where Humboldt found the magnetic equator over South America,
in 7° 1' south latitude, 1, or unity—the least intensity known is, .706, found at the
magnetic equator, over the South Atlantic, and at its most southern depression;
and it increases to 1.4 in the West Indies, and to 2.0099 upon one or more points of
the North American continent, south of the magnetic pole, and about the meridian

of 92°. That it is 1.805, at Warren, Ohio, in latitude 41° 16′, and longitude 72° 57′, and decreases to 1.774 at New Haven, Connecticut, in latitude 41° 18′. That it is but 1.348 at Paris, nearly one third less than on the same latitude in some portions of this continent. That the line of equal intensity, or "*iso-dynamic*" line, of 1^{8}⁄$_{10}$, is a closed curve of an oval shape, extending somewhat below 40°, in the longitude of Cincinnati, and reaches off nearly to Bhering's Straits, on the west; rising in a similar manner, though not so abruptly, on the east; including the great northern lakes and a considerable part of Hudson's Bay. While the iso-dynamic lines of 1^{85}⁄$_{100}$, and 1^{875}⁄$_{1000}$, are smaller ovals, included within the former. Such, at least, is the present belief from such investigations as have been made. (See an article by Professor Loomis, American Journal of Science, new series, vol. iv. p. 192.)

Our subject demands a still closer examination of the elements of magnetism and its associated electricities, and their influence upon climate and the atmosphere with a view to the solution of the questions in hand, and we will pursue the inquiry in the present chapter.

Waiving, for the present, any further notice of the fact that the counter-trades are concentrated over, and contiguous to, this area of intensity, for the purpose of examining the magnetic phenomena independently, and intending to return to a consideration of their connection with it, we observe:—That it is now well settled that the iso-geothermal lines, or lines of equal terrestrial heat, are coincident, or nearly so, with the lines of equal magnetic intensity. The points where the magnetic intensity is at a minimum, on the magnetic meridian, are the warmest points of that meridian, and those where it is most intense, the coldest.

The magnetic elements of a place may be computed from its thermal ones. The laws producing or governing the distribution of one, have an intimate physical relation with those producing or governing the other. Professor Norton ably sums up a discussion of the subject (in the American Journal of Science for September, 1847), omitting the theoretic propositions, as follows:

> "1. All the magnetic elements of any place on the earth may be deduced from the thermal elements of the same; and all the great features of the distribution of the earth's magnetism may be theoretically derived from certain prominent features in the distribution of its heat.

> "2. Of the magnetic elements, the horizontal intensity is nearly proportional to the mean temperature, as measured by Fahrenheit's thermometer; the vertical intensity is nearly proportional to the difference between the mean temperatures, at two points situated at equal distances north and south of the place, in a direction perpendicular to the iso-geothermal line; and, in general, the direction of the needle is nearly at right angles to the iso-geothermal line, while the precise course of the inflected line to which it is perpendicular may be deduced from Brewster's formula for the temperature, by differentiating and putting the differential equal to zero.

"3. As a consequence, the laws of the terrestrial distribution of the physical principles of magnetism and heat must be the same, or nearly the same; and these principles themselves must have, toward one another, the most intimate physical relations."

The magnetic elements, of which Professor Norton speaks, are the declination, dip, and horizontal and vertical forces or intensities.

I have said, that toward the areas of greatest magnetic intensity, the needle every where declines. So as intensity increases, from the magnetic equator toward the poles, the needle, when so suspended as to permit of the motion, *dips*, inclines downward, and the dip is greatest, on the same parallel, where intensity is greatest. To my mind, the magnetic elements are very intelligible. They are all attributable to attraction, and attraction is greatest where intensity is greatest. There is nothing in the earth or atmosphere to make the needle point northerly rather than in any other direction, except magnetic intensity. Thus, the greater intensity of magnetism near the northern and southern points of the globe, attracts the corresponding ends of the needle in those directions. And, as magnetism increases in quantity or intensity, and the poles are approached, the attraction increases, and the needle dips more and more, till the focus of intensity and attraction is reached, and then it becomes perpendicular. So magnetism is unequally diffused, meridionally, in or over the earth, and there are two equidistant areas where its quantity or intensity is greatest. These exert a lateral attraction upon the needle; it yields to this attraction, and hence its declination. If it is carried on to one area of intensity, and to the center of it, it will point to the northern focus of intensity or magnetic pole; and, if carried a trifle further west, it will yield to an eastern attraction, and point directly north. If carried still further west, its declination *east* will increase. Thus its normal direction is to the pole, on the central focus of intensity, and when it points directly north it is west of the central line of intensity. And thus, it seems to me, all the magnetic elements may be resolved into the one element of attraction by excess of intensity or activity.

This impression is strengthened by the fact that the needle moves to the east in the morning, when the solar rays increase magnetic activity in that direction, and west again, as their influence increases there.

Now, these elements—the declination and horizontal and vertical forces—all these periodical, regular, and irregular variations of magnetic activity, are intimately connected with the variations of atmospheric condition:

First, They show an increase of activity during certain hours of the day, corresponding to, and obviously connected with, the diurnal atmospheric changes.

Second, They show an increase of activity during the northern transit of the atmospheric machinery—an *annual* variation.

Third, They show an increase in that activity during the latter portion of each decennial period, conforming to the occurrence of solar spots.

And, fourth, *Irregular variations* of activity, corresponding with the *irregular changes* of atmospheric condition.

We will examine these results, and in doing so, take those of the element of declination—one answering for all.

The magnetic needle moves to the west in summer, from about 8 A.M. till about 2 P.M., and the extent of its progress, during that period, constitutes the magnitude of its daily variation. It is found that this variation differs in different months, and that it is normally greatest in the summer months, and least in the winter, in the ratio of about two to one. It is further found, that in different years the maximum activity occurs in different months, and that the years differ also, and there is a distinctly marked decennial period, corresponding most remarkably with the decennial maxima of recurring solar spots, as observed by Schwabe. Dr. Lamont, of Munich, gives us the following table of magnitude of declination there, for the ten years preceding 1851, which clearly exhibits this fact, and also the greater intensity during the northern transit of the atmospheric machinery. He says:

"The magnitude of the variations of declination have a period of ten years. For five years there is a uniform increase, and during the following five years a uniform decrease in the variations. With us the magnetic declination is a minimum at about eight o'clock in the morning, and is greatest at two o'clock in the afternoon. Subtracting the declination at eight o'clock from that at two o'clock, we obtain *the magnitude of the diurnal motion*. From the hourly observations, conducted in this observatory since the month of August, 1840, we ascertain the following to be the magnitude of the diurnal motion for each month separately."

	Jan.	Feb.	March.	April.	May.	June.	July.	Aug.	Sept.	Oct.	Nov.	Dec.	Autmn & Wint.	Spring & Sum.	Year.
1841	3.72	5.13	8.43	11.49	11.47	11.49	10.07	9.86	8.78	6.82	3.71	2.89	5.12	10.53	7.82
1842	3.65	4.74	8.34	10.33	9.31	9.78	8.38	9.03	7.72	7.05	3.86	2.81	5.07	9.09	7.03
1843	3.82	4.08	6.87	9.71	9.24	10.14	9.57	10.08	8.81	6.82	3.82	2.79	4.70	9.59	7.15
1844	2.81	3.43	6.95	9.53	8.42	8.88	8.38	9.28	8.23	6.54	3.94	2.98	4.44	8.79	6.61
1845	2.20	4.69	8.26	11.93	10.88	10.73	9.44	10.42	8.82	7.34	4.49	8.34	5.89	10.87	8.13
1846	3.30	6.94	9.53	12.27	12.58	11.21	11.37	11.49	10.39	7.82	5.66	3.22	6.08	11.25	8.81
1847	3.30	6.35	9.85	12.43	11.81	11.76	10.94	12.87	12.06	11.53	7.06	4.70	7.63	11.98	9.55
1848	6.52	9.01	11.96	14.56	14.22	13.80	14.67	15.40	14.00	10.30	5.78	3.53	7.85	4.44	11.05
1849	7.27	8.42	14.08	16.86	13.67	13.86	12.57	11.54	10.79	9.12	5.41	4.09	8.06	13.21	10.64
1850	5.98	8.84	12.15	14.32	14.05	13.39	12.53	12.68	12.64	9.04	6.20	3.45	7.61	13.27	10.44

The Philadelphia and Toronto observations disclose the same state of facts.

Dr. Lamont, also, in his article, gives us the following table of the magnitude of the variations derived from observations at Gottingen:

Year.	Mean of Year.
1835	9.57

1836	12.34
1837	12.27
1838	12.79
1839	11.03
1840	9.91
1841	8.70

A comparison of these tables, and particularly the latter, with Schwabe's table of spots, is interesting. There is obviously a greater mean variation when the spots are most numerous. Comparing the two with the tables of Hildreth, in relation to the temperature, from 1830 to 1840, there is, to say the least, a most remarkable coincidence. And there are others equally remarkable.

There are also irregularities of action disclosed by all, in different months of the different years, and of the same year, which are obviously connected with the difference of the seasons; and there are constantly occurring irregularities and disturbances which correspond with the, as constantly occurring, irregular atmospheric phenomena. A wide field is here opened for investigation and research. I have not time or opportunity to pursue it. Enough appears, so far as I have examined, to confirm the belief that magnetism is actively concerned in the production of the varied changes, as well as the normal conditions of the weather.

In what manner does it act? An answer to this requires an extension of the inquiry. The lines of magnetic force are every instant passing upward from the earth, *around* and *through* us. Their connection with heat is unquestionable. They are intimately associated, also, with another equally obvious and intensely active agent—electricity. We speak of this as an independent, imponderable, elementary body, but how little we yet know of it. It is every where, in every thing, easily excited into action, and then traceable to a certain, but limited extent. It is set in motion, and becomes obvious to us, by the chemical action of the acids and metals of a galvanic apparatus. We separate it from the atmosphere by friction and excitation, upon non-conductors, as in the electric machine; by the cleavage of crystals and other exciting operations. We obtain it from magnets, by the magneto-electric machine, and from the lines of magnetic force which are ever passing into the atmosphere from the earth, by intersecting them with a movable iron wire, properly insulated. *From the current of magnetism which has passed through us from the earth, electricity may thus be separated and collected over our heads.* We set it in motion, and obtain it *by heating* different metals in connection, or the same metal unequally; and from certain animals—like the torpedo and the gymnotus—whose organization is such as to enable them to evolve it. In all these cases, and they constitute an epitome of the principal methods by which we obtain it in a distinct form, it is made to flow in currents. When thus obtained, and imprisoned in non-conductors, it may be discharged, and with somewhat different effect, as it is discharged in a mass, disruptively, as it is called, as from the clouds in lightning, or permitted to flow convectively, in currents, along the wires of a galvanic apparatus, or in heated air, as from the earth to a cloud in the tornado.

It is, moreover, capable of division into positive and negative, and when concentrated or disturbed in one body, it tends to create a similar disturbance or division in a contiguous mass. To this action of electricity, the term static induction is applied. Thus, a positively electrified body *induces* a division of the electricity in a contiguous body, if both are insulated or surrounded by a non-conducting medium; the negative electricity of the contiguous body being attracted by, and tending to pass to, the positive of the adjoining body, and the positive being repelled to the opposite side. That, in its turn, if sufficiently powerful, tends to disturb the electricity of its neighbor, and attract away its negative electricity; or, if the body which contains it is free to move, to attract that. Thus, by the conflicting action of a positive atmosphere, and a negative earth, and perhaps counter-trade, influenced by magnetism and the solar rays, the currents and winds of the atmosphere are produced, the atmosphere moving with exceeding ease and rapidity. Electricity, excited into currents, or obtained and discharged in either of the methods enumerated, is identical in character, and produces certain well-known effects:

1st. Physiological.—Shocking and convulsing the animal system; producing a peculiar sensation on the tongue, and a flash before the eyes, and in sufficient quantity destroying life.

2d. Magnetic.—*Deflecting the needle*, and, by a suitable arrangement of wire into helices, *conferring magnetic power*, or constituting magnets.

3d. Luminous.—Producing light—by a spark, as it does in natural phenomena—by the glow, the brush discharge, the ball of flame, the flash, or the chain of lightning, and probably the aurora.

4th. Evolving heat.—Melting metallic substances by concentration, with a great intensity of heat—as the wire of the galvanic apparatus, and as is sometimes seen in the effects of lightning in fusing metals on persons stricken; and setting combustibles on fire.

5th. Attraction and repulsion.—Attraction, when the currents flow parallel with each other, or are of opposite natures, and repelling when of like character.

6th. Induction.—Inducing attendant circular or other secondary currents, such as may be seen in the atmosphere during its most violent displays of active energy.

7th. Capable of being dissipated by heated air, or carried off by moisture, although isolated by dry air, of ordinary temperature, which is a bad conductor.

Now, although magnetism can not be collected, imprisoned, or discharged, like electricity, or collected at all, but by its adherence to some substance capable of magnetization, it is obvious there is an intimate association, at least, between it and electricity. *They are never found alone.* All *electricity* will *magnetize*. All *magnetism* will evolve electricity. All *currents* of *electricity* have *encircling currents* of *magnetism*, and all deflect the magnetic needle. All magnetic currents give out to intersecting wires, *currents of electricity*, and all magnets *induce* them.

Electricity, therefore, whether identical in substance with magnetism, but differing in form, or whether merely associated with it, as is variously believed, should be present with magnetism in greater quantity or intensity where magnetism is most intense, and active, and whenever present, should be active and influential. And so we find, from observation, the fact to be. No inconsiderable effort has been made by the advocates of the caloric and mechanical theories, to ignore the agency of electricity and of magnetism, in the production of the varied meteorological phenomena. But it will not do. The phenomena, grouped and analyzed, disclose a potential-controlling, magneto-electric agency, and meteorology will advance rapidly to perfection, as a simple, intelligible, and practical science, *as soon as that agency is admitted.*

Electricity is always perceptibly present in storms and showers within the tropics. Most of the rain, from the tropical belt, falls from "thunder showers." So hurricanes and typhoons, and all tropical storms, are confessedly, and in proportion to their intensity, *"highly electric."* This excess of quantity or activity of electricity, exists in connection with the movable atmospheric machinery. When it moves up north in summer, and arrives at its highest point of northern transit, *storms* are very *uncommon*, and the tropical forms of cloud and showers, with thunder and lightning, prevail. This is most obvious, if not most influential, where the magnetic intensity is greatest. Violent showers, and gusts, and tornadoes, are more frequent in this country than in Europe; and over the area of greatest intensity, as in Ohio, than at a distance on the extreme eastern or western coast. And the same is true over the intense magnetic area of Asia.

Electricity, too, like magnetism, has its diurnal, and doubtless its annual and decennial variations, and also its irregular ones, and they are most obviously and intimately connected. Magnetism and electricity together, constitute the aurora. Its culmination is in the magnetic meridian—it affects the telegraph wires—is connected with the irregular disturbances which affect the magnetic needle, and does not exist in the limits of the trades, although occasionally seen from thence, when it passes south, and near them.

The aurora sometimes extends south in waves, as do the magneto-electric, atmospheric, periodical changes of cold and heat, and storm, and sunshine. *The aurora is connected with the formation of cloud,* and with a smoky atmosphere, similar to that with which we are familiar in summer and autumn. Thus Humboldt (Cosmos, vol. i. pp. 191, 192).

"This connection of the polar light with the most delicate cirrus clouds, deserves special attention, because it shows that the electro-magnetic evolution of light is a part of a meteorological process. Terrestrial magnetism here manifests its influence on the atmosphere, and on the condensation of aqueous vapor. The fleecy clouds seen in Iceland, by Thienemann, and which he considered to be the northern light, have been seen in recent times by Franklin and Richardson, near the American north pole, and by Admiral Wrangel on the Siberian coast of the Polar Sea. All remarked 'that the aurora flashed forth in the most vivid beams when masses of cirrus-strata were hovering in the upper regions of the air, and when these were so thin that their presence could only be recognized by the formation of a halo round the moon.' These clouds sometimes range themselves, even by day, in a similar manner to the beams of the aurora, and then disturb the course of the magnetic needle in the same manner as the latter. On the morning after every distinct nocturnal aurora, the same superimposed strata of clouds have still been observed that had previously been luminous. The apparently converging polar zones (streaks of clouds in the direction of the magnetic meridian), which constantly occupied my attention during my journeys on the elevated plateaux of Mexico, and in northern Asia, belong, probably, to the same group of diurnal phenomena."

Mr. William Stevenson gives us (in the London, Edinburgh, and Dublin Philosophical Magazine for July, 1853) an interesting article on the connection between aurora and clouds. His observations on this most important branch of the subject trace a connection between the aurora and the formation of cloud, and open up, as he says, "a most interesting field for observation which promises to lead to very important results." Such observations point with great significance, to the primary influence of the magneto-electricity of the earth.

To the difference in the magnetic intensity of the eastern portion of this continent, compared with Europe and our western coast, very much of the difference of climate, so far as temperature is involved, may be attributed. We have seen in what manner the iso-thermal lines surround these areas of intensity. So the most excessive climate—that is, the climate where the greatest extremes alternate, other things being equal, is upon or near the line or area of greatest magnetic intensity. I say other things being equal, because large bodies of water modify climates by equalizing the seasons—making the summers cooler and the winters warmer than the mean of the parallel.

Thus, our great interior lakes modify the climate in relation to temperature in their vicinity. Their summers are cooler and their winters warmer; but westward of them the same line of equal summer temperature, or iso-thermal line, rises with considerable abruptness, and the winter, or iso-cheimal line of equal temperature, falls in a similar manner. Thus, the range of the thermometer, from the highest elevation to the lowest depression, for the year, is very great, while in the tropics the range is comparatively small. From observations made at the military posts of the United States, Dr. Forrey deduced summer and winter lines of equal temperature, starting from the vicinity of Boston and running west, which showed most remarkably the rise of the summer lines as intensity increased, and the fall of the winter lines in like manner.

The influence of the lakes was also most obvious. The elevation of the earth increases, going west, to about 700 feet at the surface of the lakes, and to nearly 4,000 feet at the eastern base of the Rocky Mountains; and, although temperature does not decrease to as great a degree when the elevation above the level of the sea is *gradual*, yet some allowance should doubtless be made for that elevation on this line. When that allowance is made, the ascent of the summer line, to the north, over the area of greatest intensity, is strikingly apparent.

Dr. Forrey also instituted a comparison between Fort Snelling, where the climate is as excessive, and the range of the thermometer as great, as in any portion of the continent in the same latitude, with Key West, and I copy his diagram. It is very instructive, showing the gradual mean rise of the temperature, from January to December, inclusive, while the cross lines show the *extremes of each month.*

Perhaps the most interesting part of it, is the illustration of the monthly extremes, and the contrast between them, in the excessive climate of Fort Snelling, and the tropical one of Key West. Each is a type of the climate in which it is situated. The annual range and monthly extremes are small in tropical countries, and large in extra-tropical ones. The extreme range, or greatest elevation of heat, contrary to what is generally supposed, is greater at Fort Snelling than at Key West. But the climate of the latter is modified by the adjoining ocean.

I copy, also, a table (p. 304), showing the range of the thermometer for the year, and the maxima and minima, during each month, at several other places in this country, and at London and Rome, for the purpose of showing the extent of the ranges compared with those places; and also, that these great changes in each month occur very uniformly all over the country, and may always be expected, and with considerable regularity. They are incident to our climate. I wish I could engrave the foregoing diagram, and the following table, upon the mind of every man, woman, and child in the country; and under it, in ever-visible letters, these words of precaution: CONFORM TO THE PECULIARITIES OF YOUR CLIMATE, AND CLOTHE YOURSELVES, AT ALL TIMES, IN ACCORDANCE WITH THE ALTERNATIONS OF THE WEATHER. If heeded, they would save thousands, every year, from premature death.

The effect of this difference of magnetic intensity upon the climate of Europe is marked. There, the excessive summer heat, which our greater magnetic intensity and larger volume of counter trade give us, is unknown. Hence, while we can grow Indian corn (which requires the excessive summer heat) over all the Eastern States, up to 45°, and in some localities east of the lakes to 47° 30', and to 50° west of them, to the base of the Rocky Mountains, and notwithstanding the increase of elevation, they can not grow it except over a limited area, and with limited success. Nor can they, or the inhabitants of any other country except China, grow profitably the kind of cotton which is so successfully grown in the Southern States of the Union. Nor can China do so to a considerable extent, because of the mountainous character of the surface. To a level and remarkably watered country, greater magnetic and electric intensity, and a greater volume of counter-trade, we are, and ever shall remain, indebted, for an almost exclusive monopoly in the growth of two of the most important staple productions of the earth. On the other hand, although the same magnetic intensity, and its winter excess of positive electricity and cold, make our winters extreme, there are but few of the productions of temperate latitudes which we can not grow successfully, and they are comparatively unimportant.

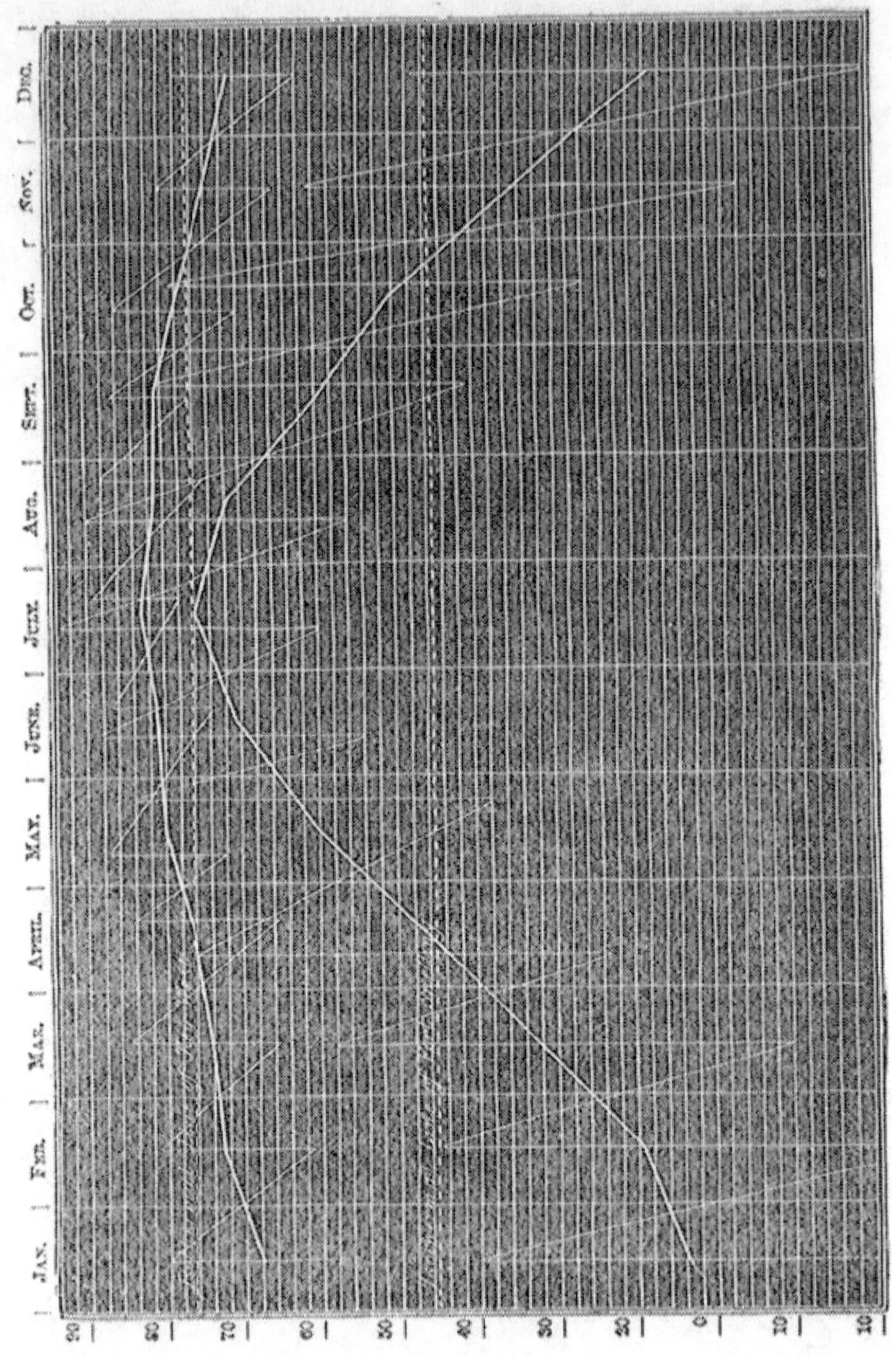

FIG. 18.

LARGER IMAGE

A. Fort Vancouver, Oregon Territory
B. Fort Brady, outlet of Lake Sup.
C. Hancock Barracks, Houlton, Me.
D. Fort Armstrong, Rock Island, Ill.
E. West Point, New York
F. Washington, D. C.
G. Jefferson Barracks, near St. Louis
H. Fort King, interior of East Florid.
I. Environs of London
J. Rome, Italy

		A.	B.	C.	D.	E.	F.	G.	H.	I.	J.
Lat.		45° 37′	46° 39′	46° 10′	41° 28′	41° 22′	38° 53′	38° 28′	29° 12′	51° 31′	41° 54′
Annual Range.		78	110	118	106	91	84	89	78	67	62
Jan.	Min.	17	-21	-24	-10	-1	14	10	33	16	29
	Max.	58	40	41	48	53	57	60	83	49	58
Feb.	Min.	32	-22	-11	-6	2	16	11	43	19	33
	Max.	55	44	42	56	56	62	70	84	54	60
Mar.	Min.	32	-7	-1	13	16	28	31	39	24	37
	Max.	60	51	54	70	72	70	76	87	60	65
Apr.	Min.	32	18	24	33	40	36	38	54	26	44
	Max.	70	62	74	78	62	73	83	93	69	74
May.	Min.	32	32	81	44	47	50	45	64	33	52
	Max.	75	79	83	84	72	85	88	97	78	80
June.	Min.	45	41	38	57	57	59	59	73	39	60
	Max.	95	86	90	89	79	92	95	105	80	88
July.	Min.	40	39	45	62	64	64	50	73	41	64
	Max.	95	84	90	95	86	94	96	102	83	91
Aug.	Min.	44	49	46	60	62	63	66	74	42	62
	Max.	95	84	85	91	87	93	96	104	79	91
Sept.	Min.	43	40	33	51	56	51	51	70	34	55
	Max.	88	75	78	87	83	88	88	99	75	85
Oct.	Min.	50	27	24	82	42	33	38	41	30	46
	Max.	66	70	72	73	69	77	80	91	68	77
Nov.	Min.	32	15	4	26	36	28	27	30	22	39
	Max.	58	58	60	64	63	66	69	82	56	67
Dec.	Min.	32	-7	-4	15	20	17	14	36	20	31
	Max.	55	42	53	62	56	61	64	79	53	60

This excess of magnetic intensity and electricity not only gives a peculiar character to our vegetation, but also to our race, our animals, and every thing. He who supposes that the restless activity and energy of the people of the United States is the result of habit, or education, or any fortuitous circumstances alone, is mistaken. Let him watch the contrast in his own feelings during those occasional languid, damp, and sultry, although not thermometrically, hot days—which so much resemble the summer weather of England—with those days of bright, bracing, N. W. and S. W. air, so much more frequent here, and he will appreciate the difference. That term "bracing," so much in use, will express the effect of this peculiar weather. It "girds up the loins," both of body and mind. Men and animals can work with more ease, even in our peculiar extremes of heat, than they can in England, and fatten with less.

A similar difference in degree is found between our climate and that of the Pacific portion of our country. Something is due to the difference in the volume and moisture of the counter-trades, and something to the contiguity of the Pacific Ocean; but to the difference in magneto-electric intensity, the contrast is mainly due. Corn and cotton will be grown, to some extent, in the valleys west of the meridian of 105°, but never as successfully as east of it.

The aurora is periodical, like all the other atmospheric phenomena, but its periodicity is not accurately ascertained. It is believed to have occurred much oftener during the second quarter of this century, than during the first. It is known, however, to occur most frequently in the spring and fall; and during those periods when the active and rapid transit of the atmospheric machinery produces the greatest degree of magnetic disturbance. This identifies it with terrestrial magnetism. Dalton gives us the following table of observations, arranged according to the months when they were seen.

	Jan.	Feb.	Mar.	Apr.	May.	June.	July.	Aug.	Sept.	Oct.	Nov.	Dec.
(1)	18	18	26	32	21	5	2	21	23	36	38	9
(2)	21	18	23	13	3	2	1	3	35	22	22	21
(3)	21	27	22	12	1	5	7	9	34	50	26	15
(4)	5	6	4	8	10	7	6	14	14	17	5	6

(1) contains those observed by him at Kendall; (2) are taken from another list; (3) is MARIAN's list of those observed before 1732; and (4), those seen in the State of New York in 1828 and 1830.

Mr. Stevenson's table of those observed by him at Dunse, from 1838 to 1847, inclusive, is as follows:

Jan.	Feb.	Mar.	Apr.	May.	June.	July.	Aug.	Sept.	Oct.	Nov.	Dec.
32	20	18	18	3	0	2	14	43	34	30	23

Observations in this country correspond substantially with the foregoing. They are, however, seen here in the summer months more frequently than in Europe. See an article by Mr. Herrick (American Journal of Science, vol. 33. p. 297). In this, also, they conform to our greater magnetic intensity and more excessive climate.

The auroras appear to follow the polar belts of condensation and precipitation. Dalton considers them indications of fair weather. They are often most brilliant just after a storm has passed, but their continuance is no indication that another will not follow within the usual period.

The condensation with which the aurora is connected, is not, in my judgment, often in the counter-trade, or below it, but above, where feeble condensation has been seen by aeronauts when invisible at the surface of the earth. Neither the height of this condensation, not that of the aurora, have been satisfactorily ascertained. The aurora of April 7th, 1847, was a favorable one for observation. It was carefully and attentively watched by Professor Olmsted, Mr. Herrick, Dr. Ellsworth, and others, and they are intelligent and skillful observers.[9] But the nature of the aurora forbids reliance on parallax, or measurements founded on the time when, any portion of the bow or arch rises in range of a particular star. The bow or arch moves southwardly, but the same rays or currents do not. The wave of magnetic *activity* moves south, and each successive current, as it is reached by the *impulse*, becomes luminous. Hence the observers, when distant, do not see, at the same time, or at different times, the same rays. The phenomenon is unquestionably magneto-electric. Electricity becomes luminous in a vacuum, and De la Rive, by combining the electric currents with those of magnetism, produced all the peculiarities of the aurora. The magnetic currents, passing from the earth, have associated electric ones in connection, and these, in the upper attenuated atmosphere, become luminous. Whether, as De La Rive supposes, by combining with the positive electricity existing there, or because the associated electric currents are *then* in excess, not being intercepted by atmospheric vapor and returned to the earth in rain, we can not know, nor is it very important we should.

Having thus taken a general view of the nature of magnetism and its associated electricities, and their connection with the general and obvious peculiarities of climate, let us approach more nearly the varied atmospheric phenomena, resulting from variations of pressure, temperature, condensation, and wind, and give them a closer consideration. They all have regularity and periodicity—they all occur in degree, and in connection with magnetism and electricity, during the twenty-four hours of every serene and normal summer's day. Grouped together, in comparison with the changes in the activity and force of the magnetic elements, their connection is clearly discernible.

The day may be said, with truth, to commence, in some portion of the summer, at 4 A.M. The atmospheric does at all seasons. At that hour the barometer is at its morning minimum. It has, as we have said, a perceptible diurnal variation of two maxima and two minima. Its periods of depression are at 4 A.M., and 4 P.M., and of elevation at 10 A.M., and 10 P.M. The difference between the elevation and depression is considerable within the tropics, where Humboldt tells us the hour of the day can be known by the height of the barometer, and it decreases toward the poles. At 4 A.M. it is then at one of its minima, and rises till 10 o'clock.

At, or about the same period, and sometimes when the barometer is falling, and previous thereto, there is a tendency to fog in localities subject to that conden-

[9] Their estimate was 100 to 120 miles.

sation. This tendency is sometimes observed at the other barometric minimum, late in the afternoon or early in the evening, but less frequently. The tendency to fog condensation is greatest in this country about the morning minimum. It seems to be owing to the influence of the earth; it is confined to the surface atmosphere, and is apparently produced by the inductive agency of the negative electricity of the earth. It disappears, whether it be high or low fog, about the time when the barometer attains its morning maximum, or about 10 A.M.

At about that period, when there has been fog, or earlier, when there has not, and sometimes as early as 8 A.M., there is a tendency to trade condensation—cirrus in mid-winter, and a cumulus in mid-summer, and, during the intermediate time, a tendency to cirro-stratus, partaking more or less of the character of one or the other, according to the season.

Temperature, in summer, commences its diurnal elevation about 4 A.M., also, and rises till about 2 P.M. From that time it falls with very little variation till 4 o'clock the next morning. It has but one maximum and one minimum in the twenty-four hours.

As the morning barometric maximum approaches, and the heat increases the magnetic activity, condensation in the trade appears, or induced condensation in the upper portion of the surface atmosphere, that portion near the earth is affected and attracted—and the "wind rises," according to the locality, the season, and the activity of the condensation. The tendency to blow increases with the tendency to trade and cumulus condensation, and continues till toward night, when it gradually dies away, unless there be a storm approaching. As the heat increases, and stimulates magnetism into activity, the magnetic needle commences moving to the west, its regular diurnal variation, and continues to do so until about 2 P.M., when it commences returning to the east, and so continues to return until 10 P.M., when it moves west again until 2 A.M., and from thence to the east, till 8 A.M.

Similar variations also take place in the horizontal force, as evinced by the action of the magnetometer needle, and in the vertical force, as shown by the oscillations. So that it is evident that there are two maxima, and two minima of magnetic activity every day, shown by all the methods by which we measure magnetic action and force—more than double at the acme of northern summer transit over that of winter, and proceeding *pari passu*, with the other daily phenomena—evincing the same irregular action which the other phenomena evince. Still another phenomenon, which has a daily change, is electric tension, or the increase or decrease in the tension of the positive or true atmospheric electricity.

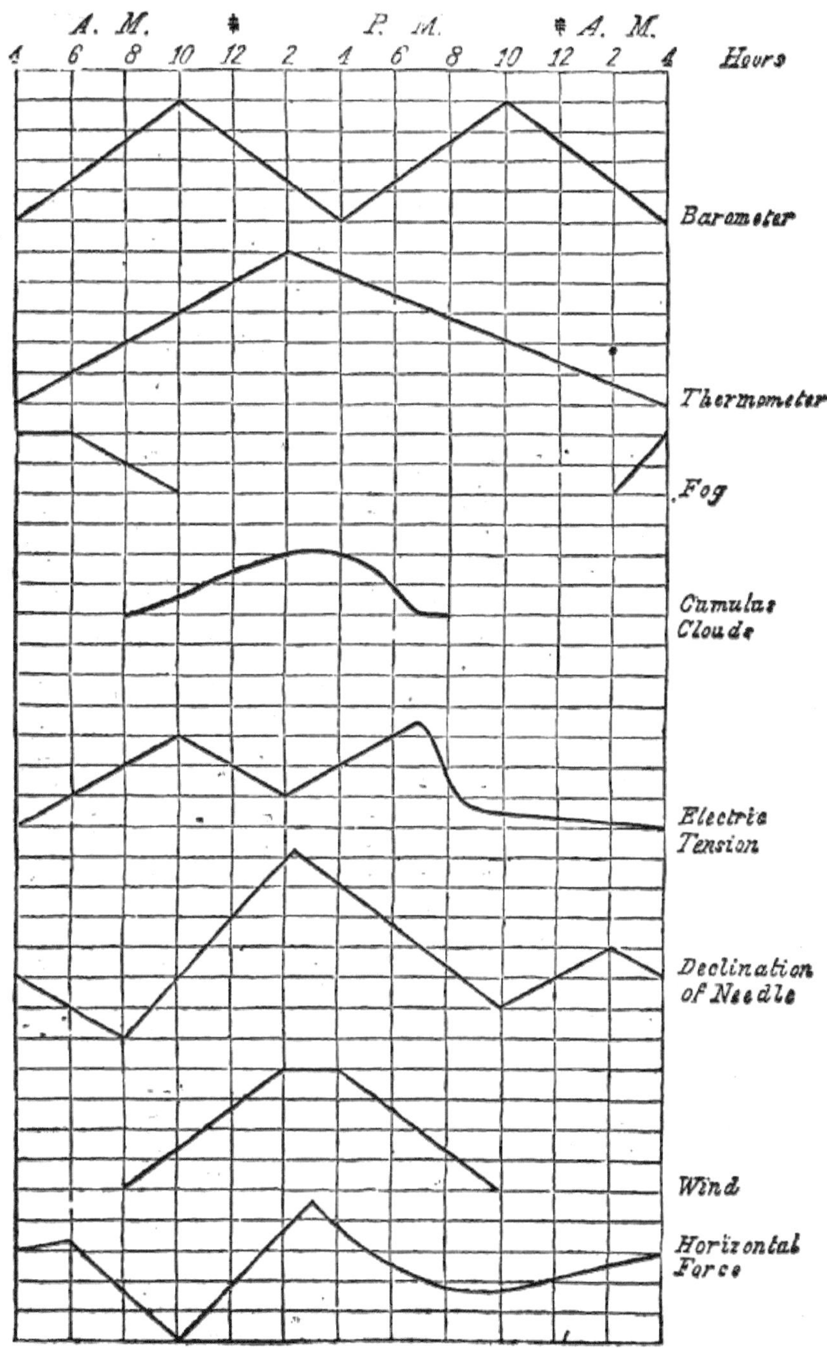

FIG. 19.

LARGER IMAGE

The following table shows the mean two hourly tensions for three years, at Kew, viz.:

Hours	12 P.M.	2 A.M.	4 A.M.	6 A.M.	8 A.M.	10 A.M.
Number of observations	655	784	804	566	1,047	1,013
Tension	22.6	20.1	20.5	34.2	68.2	88.1

Hours	12 A.M.	2 P.M.	4 P.M.	6 P.M.	8 P.M.	10 P.M.
Number of observations	848	858	878	874	878	1,007
Tension	75.4	71.5	69.1	84.8	102.4	104

From this it will be seen that the tension of electricity is at a minimum at 4 A.M., also, that it rises till 10, falls till 4 P.M., but not as rapidly, rises till 10, falls again till 4 A.M., or the close of the meteorological day—having two maxima and minima, as have most of the phenomena thus far considered.

In order to see what the connections between these ever-present, daily phenomena are, and their connection with other phenomena, and that we may understand their normal conditions, I will trace them approximately in a diagram (figure 17.)

The foregoing diagram of the daily phenomena of a summer's day, when no disturbing causes are in operation, no storm existing within influential distance, and no unusual intensity or irregular action of any of the forces present, affords a basis for considering the various phenomena of the weather in all its changes and conditions.

It is obvious that the other phenomena do not all depend upon temperature merely, if indeed any of them do.

Temperature has but one maximum and minimum, and that is exceedingly regular, and does not correspond with any other.

The barometer has two; electric tension, two; magnetic activity, two; condensation, two—one the formation of cloud, and the other the formation of fog and dew; wind, one—resembling temperature in that respect, but embracing a much less period.

Fog forms at one barometric minimum, and cloud at another.

Fog forms at one period of the magnetic variation, cloud at another.

The formation of cloud corresponds with the greatest intensity of magnetic action, and its associate electricities. But the oscillations of the barometer do not correspond with either. And thus, then, we connect them:

CAUSE.	EFFECT.	EFFECT.
Increase of magnetic or magneto-electric activity, as shown by declination and increase of horizontal and vertical force.	Decrease of pressure. Of positive electric tension. Of surface condensation, *i. e.*, fog and dew.	Increase of primary condensation. Of wind. Of electrical disturbance and phenomena in the trade and its vicinity.

This connection is equally obvious if the order is reversed—thus;

CAUSE.	EFFECT.	EFFECT.
Decrease of magnetic or magneto-electric activity.	Increase of pressure. Of tension of atmospheric electricity. Of surface condensation, *i. e.*, fog and dew.	Disappearance of primary condensation. Of wind, and Of electric disturbance in the trade and its vicinity.

If we examine still more particularly the different phenomena, we shall find the same relative action of the forces carried into all the atmospheric conditions, however violent.

1. The barometer falls when horizontal magnetic force, and a tendency to cloud and wind, increase; and rises when they decrease. This corresponds with the character of the irregular barometric oscillation. Barometric depressions accompany clouds and winds, and are in proportion to them, and are all greatest where magnetic force is greatest. The barometer also rises as the magnetic energy decreases. Do the magnetic currents, passing upward with increased force, lift, elevate the atmosphere? How, then, are we to explain the increased range of the oscillations, as the center of atmospheric machinery is reached, where magnetism has least intensity, and the perpendicular currents are less, and attraction is less? Attraction is greatest where intensity is greatest, and there the barometer stands highest, and the diurnal range is least. Is it then the attraction of magnetism which produces the barometric oscillations? If so, how then can we explain the diurnal fall while magnetism is most active?

Perhaps we have not yet arrived at such a knowledge of the nature of magnetism as is necessary to a correct answer of those questions. Faraday has taught us that the lines of magnetic force are close curves, passing into the atmosphere, and over to the opposite hemisphere, and returning through the earth, out on the opposite side in like manner, and back again, passing twice through the earth and twice through the atmosphere. All we know of this is what the iron filings indicate, and we do not know how much reliance to place upon the indications they give. But if Faraday is right, the sun will, twice each day, intersect and stimulate into increased activity the same closed magnetic curve—once when it is coming out of the earth, during our day, when its influence will be the most active, and once when it is returning on the opposite side of the earth; and a second, but feebler magnetic and electric maximum, may be occasioned by its action on the opposite and returning closed curve of the same current. However this may be, it is exceedingly difficult to conceive, of any adequate influence exerted by the tension of vapor.

So the mid-day barometric minimum may be caused by the attraction of the earth, in a state of increased magnetic activity and intensity, upon the counter-trade, and its consequent approach or settling toward the earth. Observation, as I have already said, pointedly indicates such a state of things. So the increased magnetic activity, with or by its associate electricity, acts upon the electricity of the counter-trade, condensation takes place, the electricity is disturbed in the surface-atmosphere, by induction, and its tension is changed. Opposite electrical conditions are induced in the surface strata, and attraction takes place. The air moves easily, and thus the attractions originate the winds. Secondary currents are induced, as in all other cases of electric activity, and winds, in *different strata* and directions, occur, with or without cumulus, or scud condensation, according to their activity, and the proportion of moisture of evaporation they may contain.

I am well aware that the various received theories of meteorology attribute condensation to the action of cold, mingling of colder strata, etc. But I think that view will have to be abandoned.

It assumes that moisture is evaporated and held in the atmosphere by latent heat, which is given out during condensation, and actually warms the surrounding atmosphere. Thus, the Kew Committee undertook to explain the development of greater heat, at the elevation where they, in fact, found the counter-trade. But how unphilosophical to suppose a portion of the air or vapor contained in it, can give out to another adjoining portion *more heat than is necessary to produce an equilibrium*. This can, indeed, be done by experiment—*but the experiment is made with currents of electricity*. How unphilosophical, too, to talk of latent heat in connection with evaporation, *at the lowest temperature known*. Meteorologists must revise their opinions on the subject of condensation. This latent heat has never been actually met with; on the contrary, the most sudden and complete condensations of the vapor of the atmosphere are attended by as sudden and extraordinary productions of cold, and consequent hail, and the connection between condensation and electricity is shown by too many facts to permit the old theory to stand.

Fog never forms with the thermometer below 32°. It is mainly a *summer condensation*, especially high fog. It has been attributed to the cooling effect of an atmosphere colder than the earth, but it often occurs when the earth is the coldest, and when the vapor, as it rises, is colder than the air, and could not give out heat to a warmer medium. (See American Journal of Science, vol. xliv. p. 40.) Again, it is not mere condensation, but a formation of globules or vesicles, hollow, and the air expanded in them, by means of which they float like a soap bubble which contains the warm air of the breath. Is not every vesicle a model shower, positively electrified on the outside, negatively in the center, or the reverse, according to the strata, with the air expanded in the middle by the excess of heat which negative electricity detains? Look at them, as they attach themselves to the slender nap of the cloth you wear, when passing through them, and see how many of them it would require to form a large drop of rain. The clouds are of a similar vesicular character, and rain does not fall till the vesicles unite to form drops. Sudden and extreme cold is indeed produced in the hail-storm, when, above, below, and around it, the temperature is unaffected. Testu, Wise, and other aeronauts, have so found it, and the hail tells us it is so. But it is idle to say it results from radiation. All the phenomena of the sudden, violent hail-storms are electric in an extraordinary degree. The electricity is disturbed and separated—the associated heat continues with the negative, and leaves the positive portion of the cloud, and a corresponding reduction of temperature results. So Masson found in his eudiometrical analytical experiments the *negative* wire would heat to fusion, while the positive was cold. (See London, Edinburgh, and Dublin Journal of Science for December, 1853.) This disturbed electricity is diffused over the vesicles. Listen to the thousand *crackling* sounds which initiate the clap of thunder, and may be heard when the lightning strikes near you; produced by the gathering of the lightning from as many points of the cloud where it was diffused, to unite in one current and produce the "clap" or "peal"—and to the "pouring" of the rain, which follows the union of the vesicles, after the excess of repelling electricity is discharged.

No *change* of temperature is observed when fogs form, except the ordinary change between night and day; and it seems perfectly obvious, in looking at all the phenomena, that fogs form at a temperature of 70° or 75°, in consequence of the electric influence of the earth upon the adjoining surface-atmosphere; and, when formed, they withstand the most intense action of a summer sun, till the time of day arrives for the barometric and electric tension to fall, condensation to take place in the counter-trade above, and wind to be induced. Who that has noticed the almost blistering force of the solar rays, as they break through a section of high fog, about 10 A.M., can forget them.

Fogs form near the earth, during the night, when the atmosphere above is loaded with moisture many degrees colder, and yet remains free from condensation. On the other hand, during the heat of the day, and of the hottest days, the heavy rains condense above—nay, they frequently fall at a temperature of 75° to 80°, in the tropics, and of 50° to 55° in mid-winter here.

Thus far, an adherence to the opinion that condensation was simply a cooling process; the driving out of its latent heat, not merely to another body to make an equilibrium, but *"getting rid of it"* by positive active radiation, or in some other way, so as to cool off and condense, has involved the formation and classification of clouds in obscurity. Hopkins (Atmospheric Changes, p. 331) laments this, but fettered by a false and imperfect theory, in relation to the tension of vapor, he falls into a similar error.

Now, there are, as we have seen, peculiar, distinctly-marked varieties of cloud, connected with peculiar and distinctly-marked conditions of the atmosphere, *irrespective of temperature*. None of the theories advanced, account, or profess to account for the differences in either. No modification of the calorific theory will account for them. They differ in shape, in color, in tendency to precipitation, in line of progress, and in electrical character. The explanation of this is found in the fact, that they form in distinct and different strata, partake of the positive electric character of the one, or the negative of the other; or are secondary, induced by the action of a primary condensation in a different stratum. There is not any mingling of the different strata, as has been supposed; and many other facts than those to which we have alluded, show that the formation of cloud is a magneto-electric process.

The observations of Reid show that every violent shower cloud has the electricities disturbed, and portions of it are positive, and others negative. Howard gives us the following *résumé* of Reid's observations:

"From an attentive examination of Reid's observations I have been able to deduce the following general results:

"1. *The positive electricity, common to fair weather, often yields to a negative state before rain.*

"2. *In general, the rain that first falls, after a depression of the barometer, is* NEGATIVE.

"3. *Above forty cases of rain, in one hundred, give negative* electricity; although the state of the atmosphere is positive, before and afterward.

"4. *Positive rain, in a positive atmosphere, occurs more rarely*: perhaps fifteen times in one hundred.

"5. *Snow and hail, unmixed with rain, are positive, almost without exception.*

"6. *Nearly forty cases of rain, in one hundred, affected the apparatus with both kinds* of electricity; sometimes with an interval, in which no rain fell; and so, that a positive shower was succeeded by a negative; and, *vice versâ*; at others, the two kinds alternately took place during the same shower; and, it should seem, *with a space of non-electric rain between them.*"

Howard attributes, with great apparent probability, the successive differences in the electrical character of the rain, to the passage of different portions of the cloud, having different polarity, over the place of observation. So *positive hail*, and *negative rain* fall in *parallel bands* from the same cloud. Many such instances are on record. It should be remembered that he is describing the phenomena in the showery climate of England.

But the most decisive, perhaps, as well as practically important evidence of the influence of magnetism, or magneto-electricity, in meteorological phenomena, is derived from the action of storms. My observation has been limited, for my life has been, and must be, a practical one. But, subject to future, and I hope speedy corroboration, or correction, by extensive systematic observation, I think I may venture to divide all storms into four kinds:

1. Those which come to us from the tropics, and constitute the class investigated by Mr. Redfield. That these are of a magneto-electric character is evident. They originate near the line of magnetic intensity, over, or in the vicinity of, the volcanic

islands of the tropics; are largely accompanied by electrical phenomena; extend laterally as they progress north; induce and create a change of temperature in advance of them, and do not abate until they pass off over the Atlantic to the E. or N. E., and perhaps not until they reach the Arctic circle. Their extensive and continued action is not owing to any mere *mechanical agency* of the adjoining passive air, or other supposed currents, originated, no man can tell how, but they concentrate upon themselves the local magnetic currents as they pass over and intersect them, and, by their inductive action upon the surface-atmosphere, in different directions, attract it under them, and within their more active influence. Here the action of the magnetic currents is probably the primary cause, but the power of the storm to concentrate upon itself the new magnetic currents which it intersects as it enters each new, successive field, enables them to maintain and extend their action.

The following diagram illustrates the course and gradual enlargement of a mid-autumn tropical storm, which induces a S. E. wind in front, and occasions a thaw.

FIG. 20.

2. Another class originate at the N. W., and extend gradually south easterly on the magnetic meridian. These are most frequent in summer, forming belts of showers, but occur, I believe, at all seasons of the year. They seem to be produced by magnetic waves passing south, and are followed in autumn and winter, and sometimes in summer, by the peculiar N. W. wind and scud, and a term of cooler weather.

Thus, it is believed that many, perhaps all of the alternating terms of heat and cold, are dependent on magnetic waves passing over the country in a similar manner, with a greater or less belt of condensation between them, and depending on peculiar magnetic action traveling in the same way. The S. E. extension of showers and storms, and the cooler changes of temperature which immediately follow them; with light N. W. wind in mid-summer, and with it fresher at earlier and later periods, in the form of northers blowing violently, according to the season, are intimately connected, and indicate such waves. The indication is strengthened also by the frequent progress of auroras in like manner, occurring usually after the belt of condensation has passed, and frequently following it. The clouds and currents of the atmosphere, so far as I have been able to discover, show no permanent current from the pole to the atmospheric equator, compensating for the counter-trade; and that compensation is furnished by the periodical but frequent atmospheric waves, connected with the periodical changes of storm, and cloud, and sunshine, which gradually extend from north to south, in or near the magnetic meridian. Perhaps such compensating currents are found west of the magnetic poles, as we have suggested, and make the N. E. and northerly dry winds of Western Europe and the Pacific; but, in the present state of our knowledge, it is impossible to say that they are. If it be so, the compensation they furnish must be small; for the volume of counter-trade which is not depolarized before it reaches the Arctic circle, and which passes round the magnetic pole, must be very small. A majority of our periodical changes, during the northern transit, and I believe at all seasons, are of this character; and, I have reason to believe, from observation, in one or two cases, that where belts of rains and showers begin, over *any locality* in the United States, they may assume this character. I have been in Saratoga when an easterly storm commenced *south of that place*; the condensation and mackerel sky being visible at the south, and no cloud formation or rain occurring there at the time, and have traced it afterward as a belt which had a lateral extension south-eastward. Leaving that place immediately after a belt had passed south, I have overtaken it by railroad, and run into it again before arriving at New York; and witnessed its subsequent extension south-eastwardly, out over the Atlantic. I have witnessed the approach of such a belt in the spring, at Sandusky, upon Lake Erie, and its passage over to the S. E., followed by the N. W. wind, as Mr. Bassnett describes them at Ottawa, and run under the attenuated edge of the same belt, on the same day, on the way to Pittsburg, leaving the N. W. wind behind, but finding it present again with clear sky on the following morning. I have seen hundreds of them approach from the north, and pass to S. E., out over the Atlantic; followed by the N. W. wind in spring and autumn. This class of storms pass off toward, and doubtless over the track, of our European steamers and packets. I know this, for I witness it nearly every month in the year. It is not a matter of speculation, but of actual, long-continued observation. Probably, as one approaches the Gulf Stream, and when over it, its induced winds may be more violent. It is time our navigators understood this; and that all the gales of the North Atlantic, certainly, are not rotary; and do not approach from the S. W. in the same manner as the class investigated by Mr. Redfield do. Where a fresh southerly or south-westerly wind is followed by any considerable cirro-stratus or stratus-condensation, it is usually of this character.

The following diagram exhibits the peculiarities of this class of storms. It is intended to represent the same storm or belt of showers, on *two successive* days, and, of course, its usual rate of southerly extension:

FIG. 21.

This class of storms, or belts of showers, present the following succession of phenomena in summer:

1. Still warm weather, one or more days.

2. Fresh southerly wind, one or more days; if more than one, dying away at the S. W., at night-fall, but continuing into the evening of the day before the belt of condensation arrives.

3. Belt of condensation, with or without rain or showers, with the easterly wind blowing axially, if the condensation is heavy and the belt wide; westerly if the condensation is feeble or the belt narrow—the clouds moving about E. N. E.

4. Cooler air, light N. W. in summer, heavy N. W. in autumn, winter, and spring.

And, the next period—

5. Still warm weather or light airs.

6. Southerly wind, fresh.

7. Belt of condensation.

8. Cool northerly wind.

And so on, successively, unless broken in upon by some other class.

Sometimes these periods are exceedingly regular, at other times the other classes prevail. I have much reason to believe that this is the *normal, periodic* provision for condensation of our portion of the northern hemisphere, and probably of every other where rain falls regularly in the summer season, and that the other classes are exceptions, as the hurricanes are exceptions to the normal condition of the weather every where. Perhaps in some seasons, during the northern transit, the exceptions may equal the rule, but I do not now remember such a season. In other years nearly all the storms are of this character. Thus, Dr. Hildreth (in Silliman's Journal for 1827), speaking of the year 1826, in a note to his register of that year, says: "There have been, this year, an unusual number of winds from N. or N. W. Nearly every rain the past summer has been followed with winds from the northward, when, in many previous summers, the wind continued to the southward after rain." The immediate occurrence of northerly wind after the passage of the belt of condensation, is a peculiar feature of this class of storms.

As this also will be new, and is of great practical interest, I shall be pardoned for referring to other evidence. Bermuda is in latitude 32° north. In the summer season they are within the range of the Calms of Cancer, as Lieutenant Maury terms them, and not subject to storms. From November to May, inclusive, they have successions of revolving wind. Colonel Reid gave them much attention, and studied them barometrically: that is, he studied the changes of the wind during the successive periodic depressions. He found them revolving like ours, and hence inferred the truth of the gyratory theory in relation to all winds. But it is perfectly evident the same polar belts which pass over us reach them during the southern transit. The precedent southerly wind, the *central condensation*, the appearance of lightning, and the rotation of the wind by both the east and west, but most frequently by west, are the same. In his chapter on observations at the Bermudas, he gives us many examples. Probably the existence of the Gulf Stream to the west and north has a modifying influence upon them, and their action becomes less intense in that latitude, but they are very similar. I copy a record of the weather, for a month, which may be found on pages 252, 253, and 254, and a portion of his remarks:

"The month of December, 1839, presents a continual succession of revolving winds passing over the Bermudas, with scarcely an irregularity, as regards the fall and rise of the barometer accompanying the veering of the wind. One, however, occurred on the 10th and 11th. The S. W. wind abated, and changed to W. N. W., with the barometer still falling. But in the column of remarks it is noted that there was lightning seen in the N. and N. W., from 7 P.M., during the night. This irregularity may, therefore, have been occasioned by a gale passing over the banks of Newfoundland, influencing the direction of the wind at Bermuda.

"REVOLVING WINDS.

Date.	Hour.	Direction of Wind.	Wind's Force.	Weather.	Bar.	Ther.
1839.						
Nov. 30	Midnight.	S. S. E.	1	b. c.	30·06	65
Dec. 1	Noon.	S. S. W.	3	b. c.	30·07	71
2	"	S. W.	5	g. m. q.	29·86	70
3	"	S. S. W.	3	g. c.	29·76	"
4	"	S. W.	6	g. m. r.	29·62	68
5	"	W. N. W.	5	p. q.	29·56	"
6	"	N. W.	6	p. q.	*29·55	"
7	"	N. N. W.	5	b. c.	29·78	70
"	Midnight.	N. N. W.	3	b. c.	29·89	68
8	Noon.	W. N. W.	2	b. c.	29·82	71
9	"	S. S. W.	5	p. q.	29·84	70
10	"	S. W.	2	b. c.	29·96	"
11	"	W. N. W.	6	b. c. m.	*29·88	68
12	"	S. S. W.	"	b. v.	29·99	69
13	"	N. N. by W.	"	b. v.	30·01	66
14	"	N. N. W.	5	b. c. v.	30·06	64
"	Midnight.	N. W.	2	b. c. p.	30·05	63
15	Noon.	S. W. by S.	6	g. m. r.	29·72	65
"	P.M. 2	S. S. W.	7	m. q. r.	29·92	64
"	" 4	S. S. W.	"	g. m. q. r.	29·55	"
"	" 6	W. S. W.	"	q. w.	*29·53	"
"	" 8	N. W.	6	b. c. q.	29·54	"
"	" 10	N. N. W.	"	b. c.	29·55	"
16	Noon.	N. W.	7	b. c. m.	29·53	62
17	"	N. W. by N.	"	p. q.	29·67	60

18	"	N. W.	6	c. q.	29·86	"
19	"	N. W. by N.	7	m. q. r.	*29·73	59
20	"	N. N. W.	"	p. q. c.	29·89	58
21	"	N. W. by N.	6	c. q.	29·96	56
"	Midnight.	S. W.	1	b. c.	29·95	55
22	Dawn.	— —	0			
"	Noon.	S. S. W.	5	g. m.	29·83	56
"	P.M. 4	S.	7	g. m.	29·79	"
"	" 6	S. S. E.	"	g. m. r.	29·61	"
"	" 8	S. S. E.	"	w. r.	29·52	"
"	" 10	S. E.	"	m. w. r.	29·48	"
23	Noon.	S. W.	6	b. c. m.	*29·44	57
24	"	W. N. W.	"	b. m.	29·71	59
25	"	W. N. W.	5	b. c.	29·88	56
26	"	N.	3	c.	30·09	62
27	"	S. E.	5	c. q. r.	30·07	61
28	"	S. W.	6	c. q.	29·88	66
"	Midnight.	S. S. W.	"	b. c.	29·76	65
29	Noon.	S. W.	7	c. b.	*29·48	64
30	"	W. N. W.	6	b. c. q.	29·83	55
31	"	N. W.	5	b. c.	30·12	58

"Remark printed in the Register.

"The changes of the wind during the December gales have been nearly the same in all: *i. e.,* commencing with a southerly wind at first, the wind has veered by the west, toward the northwest, sometimes ending as far round as N. N. W."

These extracts show the passage of several successive belts, each with the phenomena in regular order.

The first commences with blue sky and detached clouds, barometer up, thermometer down to 65°, and nearly calm, on the 30th of November.

Dec. 1 (at noon). Wind freshens from S. S. W.; thermometer rises; barometer still up.

Dec. 2. Barometer has fallen; thermometer up; wind increasing from S. W., with gloomy, squally appearance.

Dec. 3. Wind S. S. W.; barometer slowly falling; thermometer slightly.

Dec. 4. Wind fresh; S. W.; condensation and rain has reached them, and it carries barometer and thermometer down.

Dec. 5. Wind shifting by the west, and squally.

Dec. 6. Winds gets N. W.; blows fresh; barometer at its minimum, probably at the time of the change of wind, although the register does not show the precise time.

Dec. 7. Wind N. N. W.; blue sky and detached clouds (N. W. scud), cleared off; barometer elevated by the N. W. wind, from 29.55 to 29.78. Midnight: blue sky; detached clouds (N. W. scud probably); barometer up to 29.89; thermometer fallen, from the cooler character of the northerly wind.

Dec. 8. Wind having lulled as a northerly wind has got round to S. W. again; thermometer up; barometer falling, and another belt approaching, and so on.

The first and last part of December show each two regular occurrences of substantially the same phenomena. The middle is somewhat more irregular.

There were five distinctly-marked periods, and one squally, long-continued period, with a slight tendency to condensation, and a slight fall of barometer and rain on the 19th (N. W. squall probably), but not sufficient to reverse the wind to the south. In Colonel Reid's opinion there were five revolving gales which passed over Bermuda during the month. In my opinion, there were five perfect polar waves of condensation, and one imperfect one, with as many successive southerly winds preceding the condensation, with or without rain in the center, followed by as many cold N. W. or N. N. W. winds, with squalls, in the rear, about five days apart. (See the * in the barometric column.)

We are at issue. Let the question be determined by *actual observation*, and not by *speculation*. It is of fundamental and exceeding importance to the science.

Now, let us take a month in summer, from the observations of Mr. Bassnett, at Ottawa. Here the climate differs somewhat from that east of the Alleghanies; the magnetic intensity is greater, and the action more violent and irregular. That part of the country, it should be remembered, has a greater fall of rain in summer, for reasons we have stated, and those periodic revolutions are more frequent.

> "A brief abstract from a journal of the weather for one sidereal period of the moon, in 1853.
>
> "*June* 21st. Fine clear morning (S. fresh): noon very warm 88°; 4 P.M., plumous *cirri in south*; ends clear.
>
> "22d. Hazy morning (S. very fresh) arch of cirrus in west; 2 P.M., black in W. N. W.; 3 P.M., overcast and rainy; 4 P.M., a heavy gust from south; 4.30 P.M., blowing furiously (S. by W.); 5 P.M., tremendous squall, uprooting trees and scattering chimneys; 6 P.M., more moderate (W.).
>
> "23d. Clearing up (N. W.); 8 A.M., quite clear; 11 A.M., bands of mottled cirri pointing N. E. and S. W., ends cold (W. N. W.); the cirri seem to rotate from left to right, or with the sun.
>
> "24th. Fine clear, cool day, begins and ends (N. W.).
>
> "25th. Clear morning (N. W. light); 2 P.M. (E.), calm; tufts of tangled cirri in north, intermixed with radiating streaks, all pass-

ing eastward; ends clear.

"26th. Hazy morning (S. E.), cloudy; noon, a heavy, windy-looking bank in north (S. fresh), with dense cirrus fringe above, on its upper edge; clear in S.

"27th. Clear, warm (W.); bank in north; noon bank covered all the northern sky, and fresh breeze; 10 P.M., a few flashes to the northward.

"28th. Uniform dense cirro-stratus (S. fresh); noon showers all round; 2 P.M., a heavy squall of wind, with thunder and rain (S. W. to N. W.); 8 P.M., a line of heavy cumuli in south; 8.30 P.M., a very bright and high cumulus in S. W., protruding through a layer of dark stratus; 8.50 P.M., the cloud bearing E. by S., with three rays of electric light.

"29th. A stationary stratus over all (S. W. light); clear at night, but distant lightning in S.

"30th. Stratus clouds (N. E. almost calm); 8 A.M., raining gently; 3 P.M., stratus passing off to S.; 8 P.M., clear, pleasant.

"*July* 1st. Fine and clear; 8 A.M., cirrus in sheets, curls, wisps, and gauzy wreaths, with patches beneath of darker shade, all nearly motionless; close and warm (N. E.); a long, low bank of haze in S., with one large cumulus in S. W., but very distant.

"2d. At 5 A.M., overcast generally, with hazy clouds and fog of prismatic shades, chiefly greenish-yellow; 7 A.M. (S. S. E. freshening), thick in W.; 8 A.M. (S. fresh), much cirrus, thick and gloomy; 9 A.M., a clap of thunder, and clouds hurrying to N.; a reddish haze all around; at noon the margin of a line of yellowish-red cumuli just visible above a gloomy-looking bank of haze in N. N. W. (S. very fresh); warm, 86°; more cumuli in N. W.; the whole line of cumuli N. are separated from the clouds south by a clearer space. These clouds are borne rapidly past the zenith, but never get into the clear space—they seem to melt or to be turned off N. E. The cumuli in N. and N. W., slowly spreading E. and S.; 3 P.M., the bank hidden by small cumuli; 4 P.M., very thick in north, magnificent cumuli visible sometimes through the breaks, and beyond them a dark, watery back-ground (S. strong); 4.30 P.M., wind round to N. W. in a severe squall; 5 P.M., heavy rain, with thunder, etc.—all this time there is a bright sky in the south visible through the rain 15° high; 7 P.M., clearing (S. W. mod.).

"3d. Very fine and clear (N. W.); noon, a line of large cumuli in N., and dark lines of stratus below, the cumuli moving eastward; 6 P.M., their altitude 2° 40'. Velocity, 1° per minute; 9 P.M., much lightning in the bank north.

"4th. 6 A.M., a line of small cumulo-stratus, extending east and west, with a clear horizon north and south 10° high. This band seems to have been thrown off by the central yesterday, as it moves slowly south, preserving its parallelism, although the clouds composing it move eastward. Fine and cool all day (N. W. mod.)—lightning in N.

"5th. Cloudy (N. almost calm), thick in E., clear in W.; same all day.

"6th. Fine and clear (E. light); small cumuli at noon; clear night.

"7th. Warm (S. E. light); cirrus bank N. W.; noon (S.) thickening in N.; 6 P.M., hazy but fine; 8 P.M., lightning in N.; 10 P.M., the lightning shows a heavy line of cumuli along the northern horizon; calm and very dark, and incessant lightning in N.

"8th. Last night after midnight commencing raining, slowly and steadily, but leaving a line of lighter sky south; much lightning all night, but little thunder.

"8th. 6 A.M., very low scud (500 feet high) driving south, still calm below (N. light); 10 A.M., clearing a little; a bank north, with cirrus spreading south; same all day; 9 P.M., wind freshening (N. stormy); heavy cumuli visible in S.; 10.30 P.M., quite clear, but a dense watery haze obscuring the stars; 12 P.M., again overcast; much lightning in S. and N. W.

"9th. Last night (2 A.M. of 9th) squall from N. W. very black; 4 A.M., still raining and blowing hard, the sky a perfect blaze, but very few flashes reach the ground; 7 A.M., raining hard; 8 A.M. (N. W. strong); a constant roll of thunder; noon (N. E.); 2 P.M. (N.); 4 P.M., clearing; 8 P.M., a line of heavy cumuli in S., but clear in N. W., N., and N. E.

"10th. 3 A.M., Overcast, and much lightning in south (N. mod.); 7 A.M., clear except in south; 6 P.M. (E.); 10 P.M., lightning south; 11 P.M., auroral rays long, but faint, converging to a point between Epsilon Virginis and Denebola, in west; low down in west, thick with haze; on the north the rays converged to a point still lower; lightning still visible in south. This is an aurora in the west.

"11th. Fine, clear morning (N. E.); same all day; no lightning visible to-night, but a bank of clouds low down in south, 2° high, and streaks of dark stratus below the upper margin.

"12th. Fine and clear (N. E.); noon, a well-defined arch in S. W., rising slowly; the bank yellowish, with prismatic shades of greenish-yellow on its borders. This is the O. A. At 6 P.M., the bank spreading to the northward. At 9 P.M., thick bank of haze in north, with bright auroral margin; one heavy pyramid of light passed through Cassiopeia, traveling *westward* 1½° per minute. This moves to the other side of the pole, but not more inclined toward it than is due to prospective, if the shaft is very long; 11.10 P.M., saw a mass of light more diffuse due east, reaching to *Markab*, then on the prime vertical. It appears evident this is seen in profile, as it inclines downward at an angle of 10° or 12° from the perpendicular. It does not seem very distant. 12 P.M., the aurora still bright, but the brightest part is now west of the pole, before it was east.

"13th. 6 A.M., clear, east and north; bank of cirrus in N. W., *i. e.*, from N. N. E. to W. by S.; irregular branches of cirrus clouds, reaching almost to south-eastern horizon; wind changed (S. E. fresh); 8 A.M., the sky a perfect picture; heavy regular shafts of dense cirrus radiating all around, and diverging from a thick nucleus in north-west, the spaces between being of clear, blue sky. The shafts are rotating from north to south, the nucleus advancing eastward.

"At noon (same day), getting thicker (S. E. very fresh); 6 P.M., moon on meridian, a prismatic gloom in south, and very thick stratus of all shades; 9 P.M., very gloomy; wind stronger (S. E.); 10 P.M., very black in south, and overcast generally.

"14th. Last night, above 12 P.M., commenced raining; 3 A.M., rained steadily; 7 A.M., same weather; 8.20 A.M., a line of low storm-cloud, or scud, showing very sharp and white on the dark back-ground all along the southern sky. This line continues until noon, about 10° at the highest, showing the northern boundary of the storm to the southward; 8 P.M., same bank visible, although in rapid motion eastward; same time clear overhead, with cirrus fringe pointing north from the bank; much lightning in south (W. fresh); so ends.

"15th. Last night a black squall from N. W. passed south without rain; at 3 A.M., clear above but, very black in south (calm below all the time); 9 A.M., the bank in south again throwing off rays of cirri in a well-defined arch, whose vortex is south; these pass east, but continue to form and preserve their linear direction to the north; no lightning in south to-night.

"16th. Clear all day, without a stain, and calm.

"17th. Fine and clear (N. E. light); 6 P.M., calm.

"18th. Fair and cloudy (N. E. light); 6 P.M., calm.

"19th. Fine and clear (N. fresh); I. V. visible in S. W.

"20th. 8 A.M., bank in N. W., with beautiful cirrus radiations; 10 A.M., getting thick, with dense plates of cream-colored cirrus visible through the breaks; gloomy looking all day (N. E. light)."

The letters in a parenthesis signify the direction of the wind.

During this month there were three distinctly marked periods of belts of showers, preceded by "fresh" or "strong" south wind, and followed by the N. W. There was a period when a belt of less intense stratus, without much wind, occurred (28th, 29th, and 30th of June). This was followed by a distinct belt of showers and *fresh* S. wind, on the 2d of July, and by the N. W. wind and clear weather, on the 3d.

During the rest of July it was more irregular, with the exception of the 7th, 8th, and 9th, when another belt and revolution occurred.

Now, these periods, when distinctly marked, exhibit the same succession of phenomena—viz., elevation of temperature, fresh southerly wind, belt of condensation, cumulus or stratus with cirrus running east, but extending south, followed by N. W. wind, and clear, cold air. Can any one believe they were successive rotary gales?

I wish, in this connection, to make a suggestion to Lieutenant Maury and others. The descriptions of M. Bassnett, although not perfect, are very intelligible. He describes things as they were, and as they should be described. He distinguishes the clouds, and the scud, and other appearances.

But Colonel Reid's descriptions are unmeaning and unintelligible. G. M.— Gloomy, misty! Gloomy from what? fog, or stratus, or a stratum of scud, or what? We can not know. Again, C. The table tells us this stands for detached clouds. But of what kind? Cumulus, broken stratus, patches of cirro-cumulus or cirro-stratus, or scud? All these, and indeed every kind of cloud or fog formation, except low fog, may exist in detached portions.

These abbreviations will not answer; they do not describe the weather. The clouds must be studied and described. There is no difficulty in doing it. Sailors will learn them very soon after their teachers have; and those who teach them should see to it that the logs contain terms of description which convey the meaning which may, and ought to be, conveyed. The use of these indefinite terms can not be continued without culpability.

Again, the observations of seamen off our coast are in accordance with the progress of this class of storms on land, and prove that they continue S. E. over the Atlantic, abating in action as they approach the tropics. There is abundant evidence of this in the work of Colonel Reid, and the charts of Lieutenant Maury, but I can not devote further space to them.

The third class form in the counter-trade, over some portion of the country, from excessive volume or action of the counter-trade, or local magnetic activity, without coming from the tropics or being connected with a regular polar wave of magnetic disturbance.

The following diagram exhibits their form, progress, and accompanying induced winds.

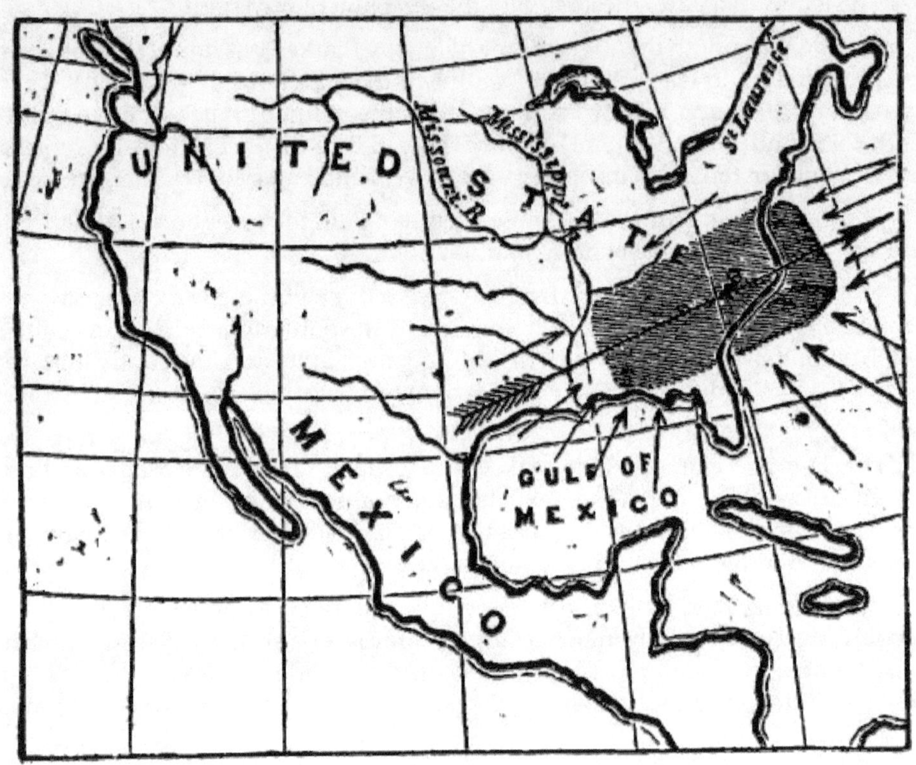

FIG. 22.

The gentle rains of spring, particularly April, and the moderate and frequent snow-storms of winter, are often of this character; and so are the heavy rains, which commence at the morning barometric minimum, rain heavily through the forenoon, and light up near mid-day in the south, followed by gentle, warm, S. W. winds. This class are more frequent in some years than others—probably the early years of the decade, while polar storms are, during the later ones. It is this class which have *violent* easterly winds *in front*, and on the *south side*, with two or more currents, and which Mr. Redfield has also supposed to be cyclones.

The fourth class are isolated showers, occurring over particular localities, or belts of drought and showers alternating; sometimes a general disposition to cloudy and showery weather for a longer or shorter interval over the whole country; at others, limited to particular localities in the course of the trade. Such a period occurred during the wheat harvest of 1855. This class I attribute to a general increased magnetic action, but it may be induced by an increased volume, or greater south polar magnetic intensity of the counter-trade, exciting and concentrating the regular currents of the field, and increasing their activity and energy. These also often work off south gradually, and are followed by a cold N. W. air for a day or two; showing a tendency, in the excited magnetism, to pass as a wave toward the tropics.

The following diagram will give some idea of this class:

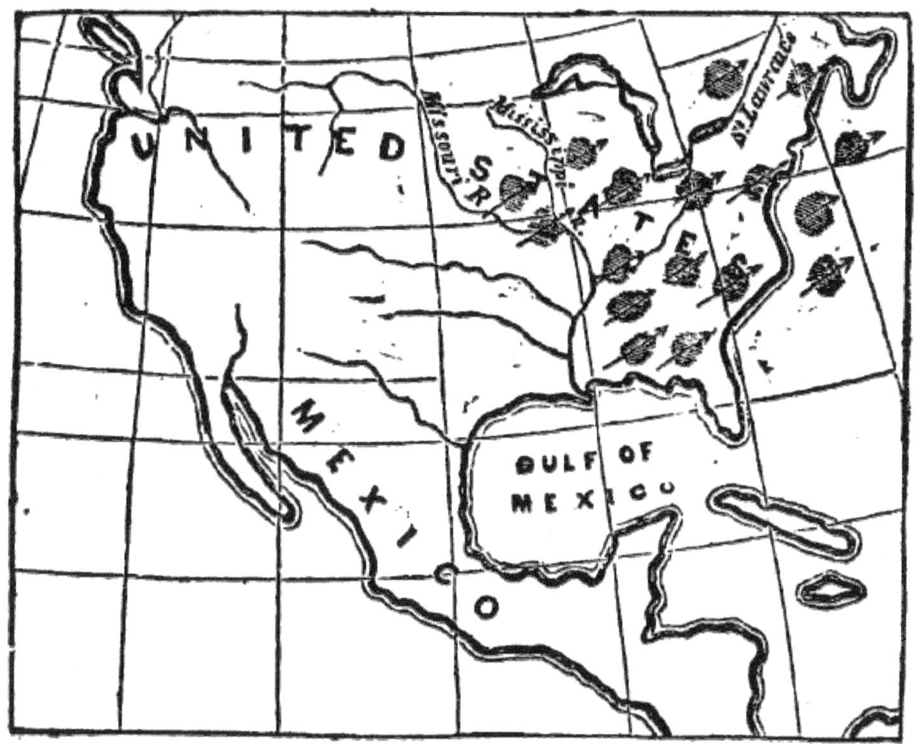

FIG. 23.

There are sometimes very obvious local tendencies to precipitation over portions adjoining an area affected with drought, as there are other magnetic irregularities over particular areas.

All these classes of storms are variant in intensity. Sometimes the general or local cloud-formation is weak, and does not produce precipitation at all; so of that which extends southerly. Probably the tropical storm are always sufficiently dense and active to precipitate. Their action is often violent over particular localities, and hence the more frequent occurrence of the tornado over the more intense area of Ohio, and other portions of the west. All violent local storms are doubtless owing to local magneto-electric activity.

CHAPTER XI.

Prognostics

The reader who has attentively perused and considered the facts stated, and the principles deduced, in the preceding pages, and is ready to make a practical application of them by careful observation, will have little difficulty in understanding the varied atmospheric conditions; and will soon be able to form a correct judgment of the immediate future of the weather, so far as his limited horizon will permit.

But there are other facts and considerations, not specifically alluded to, which will materially aid him in his observations; and there is a degree of philosophical truth in the proverbs and signs, which ancient popular observation accumulated, and poetry and tradition have preserved, that meteorologists have been slow to discover or admit, but which will be obvious upon examination, and commend them to his attention.

The classical reader is doubtless familiar with that part of the first Georgic of Virgil, which contains a description of the signs indicative of atmospheric changes. Much of it is beautifully poetic, and, if read in the light of a correct philosophy, is equally truthful.

I copy from a creditable translation, found in the first volume of Howard's "Climate of London":

> "All that the genial year successive brings,
> Showers, and the reign of heat, and freezing gales,
> Appointed signs foreshow; the Sire of all
> Decreed what signs the southern blast should bring,
> Decreed the omens of the varying moon:
> That hinds, observant of the approaching storm,
> Might tend their herds more near the sheltering stall."

PROGNOSTICS.—1st. Of Wind.

> "When storms are brooding—in the leeward gulf
> Dash the swell'd waves; the mighty mountains pour
> A harsh, dull murmur; far along the beach
> Rolls the deep rushing roar; the whispering grove
> Betrays the gathering elemental strife.
> Scarce will the billows spare the curved keel;
> For swift from open sea the cormorants sweep,
> With clamorous croak; the ocean-dwelling coot

Sports on the sand; the hern her marshy haunts
Deserting, soars the lofty clouds above;
And oft, when gales impend, the gliding star
Nightly descends athwart the spangled gloom,
And leaves its fire-wake glowing white behind.
Light chaff and leaflets flitting fill the air,
And sportive feathers circle on the lake."

2d. Of Rain.

"But when grim Boreas thunders; when the East
And black-winged West, roll out the sonorous peal,
The teeming dikes o'erflow the wide champaign,
And seamen furl their dripping sails. The shower,
Forsooth, ne'er took the traveler unawares!
The soaring cranes descried it in the vale,
And shunn'd its coming; heifers gazed aloft,
With nostrils wide, drinking the fragrant gale;
Skimm'd the sagacious swallow round the lake,
And croaking frogs renew'd their old complaint.
Oft, too, the ant, from secret chambers, bears
Her eggs—a cherished treasure—o'er the sand,
Along the narrow track her steps have worn.
High vaults the thirsty bow; in wide array
The clamorous rooks from every pasture rise
With serried wings. The varied sea-fowl tribes,
And those that in Cäyster's meadows seek,
Amid the marshy pools, their skulking prey,
Fling the cool plenteous shower upon their wings,
Crouch to the coming wave, sail on its crest,
And idly wash their purity of plume.
The audacious crow, with loud voice, hails the rain
A lonesome wanderer on the thirsty sand.
Maidens that nightly toil the tangled fleece,
Divine the coming tempest; in the lamp
Crackles the oil; the gathering wick grows dim."

3d. Of Fair Weather.

"Nor less, by sure prognostics, mayest thou learn
(When rain prevails), in prospect to behold
Warm suns, and cloudless heavens, around thee smile.
Brightly the stars shine forth; Cynthia no more
Glimmers obnoxious to her brother's rays;
Nor fleecy clouds float lightly through the sky.
The chosen birds of Thetis, halcyons, now
Spread not their pinions on the sun-bright shore;
Nor swine the bands unloose, and toss the straw.
The clouds, descending, settle on the plain;

While owls forget to chant their evening song,
But watch the sunset from the topmost ridge.
The merlin swims the liquid sky, sublime,
While for the purple lock the lark atones:
Where she, with light wing, cleaves the yielding air,
Her shrieking fell pursuer follows fierce—
The dreaded merlin; where the merlin soars,
Her fugitive swift pinion cleaves the air.
And now, from throat compressed, the rook emits,
Treble or fourfold, his clear, piercing cry;
While oft amid their high and leafy roosts,
Bursts the responsive note from all the clan,
Thrill'd with unwonted rapture—oh! 'tis sweet,
When bright'ning hours allow, to seek again
Their tiny offspring, and their dulcet homes.
Yet deem I not, that heaven on them bestows
Foresight, or mind above their lowly fate;
But rather when the changeful climate veers,
Obsequious to the humor of the sky;
When the damp South condenses what was rare,
The dense relaxing—or the stringent North
Rolls back the genial showers, and rules in turn,
The varying impulse fluctuates in their breast:
Hence the full concert in the sprightly mead—
The bounding flock—the rook's exulting cry."

4th. The Moon's Aspects, etc.

"Mark with attentive eye, the rapid sun—
The varying moon that rolls its monthly round;
So shalt thou count, not vainly, on the morn;
So the bland aspect of the tranquil night
Will ne'er beguile thee with insidious calm.
When Luna first her scatter'd fires recalls,
If with blunt horns she holds the dusky air,
Seamen and swains predict th' abundant shower.
If rosy blushes tinge her maiden cheek,
Wind will arise: the golden Phœbe still
Glows with the wind. If (mark the ominous hour!)
The clear fourth night her lucid disk define,
That day, and all that thence successive spring,
E'en to the finished month, are calm and dry;
And grateful mariners redeem their vows
To Glaucus, Inöus, or the Nereid nymph."

5th. The Sun's Aspects, etc.

"The sun, too, rising, and at that still hour,
When sinks his tranquil beauty in the main,

Will give thee tokens; certain tokens all,
Both those that morning brings, and balmy eve.
When cloudy storms deform the rising orb,
Or streaks of vapor in the midst bisect,
Beware of showers, for then the blasting South
(Foe to the groves, to harvests, and the flock),
Urges, with turbid pressure, from above.
But when, beneath the dawn, red-fingered rays
Through the dense band of clouds diverging, break,
When springs Aurora, pale, from saffron couch,
Ill does the leaf defend the mellowing grape;
Leaps on the noisy roof the plenteous hail,
Fearfully crackling. Nor forget to note,
When Sol departs, his mighty day-task done,
How varied hues oft wander on his brow;
Azure betokens rain: the fiery tint
Is Eurus's herald; if the ruddy blaze
Be dimm'd with spots, then all will wildly rage
With squalls and driving showers: on that fell night,
None shall persuade me on the deep to urge
My perilous course, or quit the sheltering pier.
But if, when day returns, or when retires,
Bright is the orb, then fear no coming rain:
Clear northern airs will fan the quiv'ring grove.
Lastly, the sun will teach th' observant eye
What vesper's hour shall bring; what clearing wind
Shall waft the clouds slow floating—what the South
Broods in his humid breast. Who dare belie
The constant sun?"

I copy also the following from Howard:

"Dr. Jenner's signs of rain—an excuse for not accepting the invitation of a friend to make a *country* excursion.

"The hollow winds begin to blow,
The clouds look black, the glass is low,
The soot falls down, the spaniels sleep,
And spiders from their cobwebs creep.
Last night the sun went pale to bed,
The moon in halos hid her head,
The boding shepherd heaves a sigh,
For see! a rainbow spans the sky.
The walls are damp, the ditches smell;
Closed is the pink-eyed pimpernel.
Hark! how the chairs and tables crack;
Old Betty's joints are on the rack.
Loud quack the ducks, the peacocks cry;

> The distant hills are looking nigh.
> How restless are the snorting swine!—
> The busy flies disturb the kine.
> Low o'er the grass the swallow wings;
> The cricket, too, how loud it sings!
> Puss, on the hearth, with velvet paws,
> Sits smoothing o'er her whisker'd jaws.
> Through the clear stream the fishes rise
> And nimbly catch the incautious flies;
> The sheep were seen, at early light,
> Cropping the meads with eager bite.
> Though *June*, the air is cold and chill;
> The mellow blackbird's voice is still;
> The glow-worms, numerous and bright,
> Illumed the dewy dell last night;
> At dusk the squalid toad was seen,
> Hopping, crawling, o'er the green.
> The frog has lost his yellow vest,
> And in a dingy suit is dress'd.
> The leech, disturbed, is newly risen
> Quite to the summit of his prison.
> The whirling wind the dust obey
> And in the rapid eddy plays.
> My dog, so altered in his taste,
> Quits mutton-bones, on grass to feast;
> And see yon rooks, how odd their flight!
> They imitate the gliding kite:
> Or seem precipitate to fall,
> As if they felt the piercing ball.
> 'Twill surely rain; I see, with sorrow,
> Our jaunt must be put off to-morrow."

Howard attributes the foregoing to Jenner; but Hone, in his "Every-Day Book," attributes it to Darwin, and gives it, with several couplets, not found in that attributed to Jenner. These I add from Hone, as follows:

> "Her corns with shooting pains torment her—
> And to her bed untimely send her."

That couplet is included by Hone with what is said of Aunt Betty.

> "The smoke from chimneys right ascends,
> Then spreading back to earth it bends.
> The wind unsteady veers around;
> Or, settling in the south is found."

Those are as philosophically accurate and valuable as any.

> "The tender colts on back do lie;
> Nor heed the traveler passing by.

In fiery red the sun doth rise,
Then wades through clouds to mount the skies."

The first of those couplets is untrue. It is doubtless alluded to as one of the acts of the animal creation, indicating sleepiness and inaction, which precede storms; but colts do not lie on the back. The other couplet is both true and important. This collection entire, whether written by Darwin or Jenner, contains most of the signs which have been preserved, and which are of much practical importance in our climate.

It is unquestionably true that "appointed signs foreshow the weather," to a great extent, every where, but with more certainty in the climate in which Virgil wrote than in our variable and excessive one. "Showers" and "freezing gales" we can, perhaps, as well understand; but the *"reign of heat,"* by which he probably meant the dry period, when the southern edge of the extra-tropical belt of rains is carried up to the north of them, we do not experience. Something like it we did indeed have, during the excessive northern transit, in the summer of 1854; but it was an exception, not the rule.

Some of the most important of those signs from Virgil and Jenner I propose to allude to in detail; but it is necessary to look; in the first place, to the character of the season and the month.

We have seen that the years differ during different periods of the same decade. That they incline to be hot and irregular during the early part of it, and cool, regular, and productive during the latter portion—subject, however, to occasional exceptions. The latter half of the third decade of this century (1826 to 1830, inclusive) was comparatively warm; and, in the latitude of 41°, was very unhealthy, and so continued during the early part of the next, over the hemisphere, embracing the *cholera seasons*. The spots upon the sun were much less numerous than usual, during the latter half of the third decade. Thus the spots from

1826 to 1830, inclusive, were 873

1836 to 1840, inclusive, were 1201

1846 to 1850, inclusive, were 1168

and the size of those from 1836 to 1840 exceeded those of the other years.

The attentive observer will very soon be satisfied that the seasons have a character; and those of every year differ in a greater or less degree from those of other years in the same decade, and those of one decade not unfrequently from those of some other. *Periodicity* is stamped upon all of them, and upon all resulting consequences. Like seasons come round, and, like productiveness or unproductiveness, healthy or epidemic diatheses, attend them. We have seen that, in relation to mean temperature, there are such periodical diversities, but they are more strongly marked in the character of storms, and other successions of phenomena. *"All signs fail in a drouth,"* for then all attempts at condensation are partial, imperfect, and ineffectual. *"It rains very easy,"* it is said, at other times, and so it seems to do, and with comparatively little condensation. In the one case, no great reliance can be placed upon indications which are entirely reliable in the other. So *"all our storms clear off cold,"* or, *"all our storms clear off warm,"* are equally common expressions—as the *prevailing classes* of storms give a *character* to the *seasons*. It *"rains every Sunday now,"* is sometimes said, and is often peculiarly true—the storm waves having just then a weekly or semi-weekly period, and one falls upon Sunday for several successive weeks; and when it is so, *that* coincidence is sure to be noticed and commented upon, and the other perhaps disregarded.

If the seasons depended upon the northward and southward journey of the sun alone, entire regularity might be expected—for we have no reason to believe that magnetism and electricity contain, within themselves, inherently, any tendency to irregularity, or periodicity; and, the sun being constant in his *periods*, would be constant in his *influence*. But he is inconstant and variable in his influence, and it is apparently traceable to the existence of spots; but I am not quite sure that it is occasioned by the *observable* spots alone. Grant that the intensity and power of his rays differ on the same day, in different years, and that difference may be attributable in part to causes which our telescopes can not discover.

But the differences in the seasons do not depend on the variability of the sun's influence alone. This appears from the frequent meridional and latitudinal diversities and contrasts, to which allusion has been made. The sun can not be supposed to exert a *less* influence on a middle, than a more northern latitude; nor on one series of meridians, than another. There must, therefore, be another local and powerful disturbing cause, varying the magnetic and electric activity and influence upon the trades, as well in their incipiency as in their circuits, and thus controlling the atmospheric conditions locally and in *the opposite hemispheres*. That other disturbing cause is *volcanic action*. We can conceive of none other, and we can detect and trace the influence of that to a considerable extent. Unfortunately we know, and can practically know, comparatively little of it. It has been busy with the earth since the creation, and will continue to be so till, possibly, by a collision, it shall burst into asteroids—its molten interior flowing out in seeming combustion—each fragment retaining its magnetic polarities entire, and continuing on in an independent orbit in the heavens, an asteroid, or meteorite.

While, therefore, the agency of magnetism in itself may be regular, and the transit of the sun is regular, and "seed-time and harvest shall not cease," yet the sun is not regular in his influence, and the magnetic agency is disturbed by another and irregular power. And, although we can trace the influence of both upon the seasons, we can not measure that influence, and from it reliably foretell the weather. The discoveries of Swabe, and future ones, relative to solar irregularities, will assist us, but, till we understand better, and to some extent anticipate, the changes of volcanic action, we shall not be able to understand or foresee all the differences in the seasons. That time may come; for progress is yet to be read in the front of meteorology, and simultaneous practical observations made and interchanged at every important point on the globe. Nevertheless, the seasons have a character—often a regular one—one class of storms prevailing over all others—one series of phenomena occurring to the exclusion of others—and we must regard it if we would arrive at intelligent estimates of their future condition.

The most difficult part to understand are the meridional contrasts. Last year we had one of the worst drouths which has occurred since the settlement of the country. But while all the eastern portion of the United States was dry, New Mexico was unusually wet; and the North-western States, on the same curving line of the counter-trade, were not affected by the drouth.

Extract from a letter written by Governor Merriweather, to Mr. Bennett, in answer to a circular, published in the "New York Herald," and dated

"Santa Fe, New Mexico, Oct. 25th, 1854.

"More rain has fallen during the last six months, within this territory, than ever was known to have fallen in the same length of time, in this usually dry climate. Generally, little or no crops have been produced without irrigation; but this season some good crops have been produced without any artificial watering."

We have seen that there was an apparent connection between the remarkable volcanic action, exerted beneath the western continents during the second decade of this century, and the remarkable coldness of that decade. And it is easy to see that the comparative absence of volcanic action from immediately beneath the Old World, and its presence in great excess beneath the New, may disturb the regular action of terrestrial magnetism above it in the earth's-crust here, and affect seasons, diatheses, and health unfavorably; while from its absence they may be favorably affected there. I have some general views in relation to this, but they are necessarily speculative, for the data are few, and I reserve them.

I am, however, induced to believe that the transit of the atmospheric machinery is greater over some portions of the northern hemisphere, in some seasons, than others. The most natural explanation of the unusual contrast between the drouth of the Eastern States, and the wet of the Territories, during the last summer, is, that the concentrated counter-trade was carried west, by some irregular magnetic action in the South Atlantic or West Indies. But there was much evidence that the northern extension of the atmospheric machinery was greater than usual. The transit began *early*—it was evidently *rapid*; the rains of May fell in April, and the spring was wet; *summer set in earlier*—all the appearances then were unusually tropical—the polar belts of condensation descended upon us, but they were feeble, as they doubtless become, when they reach the tropics, and did not precipitate; the summer continued full twenty days later—no rain falling till about the 10th of September. The season throughout was excessive, but otherwise regular. Spring came earlier; summer commenced earlier and continued longer; autumn held off later, and cold weather, when it came, was uniform and severe. This season the transit has seemed to be less than for several years.[10] The spring was backward; the summer cool, but exceedingly regular; the autumn thus far without extremes, and the whole year healthy and productive. It is the normal period of the decade, between the irregular heat of the first part, and the irregular cold of the last; and it has been normal in character, and conformed beautifully to its location. If the tran-

[10] Since the text was in type, and, as might have been anticipated, we have intelligence confirmatory of this, from the Cape De Verde Islands. The inter-tropical belt of rains has not moved as far north as the northern islands—they have had no rain—and the people are in a starving condition.

sit of 1854 was further north than the mean, as it seemed to be over this country, that of itself would convey the showers which follow up in the western portion of the concentrated trade, on the east of the mountains of Mexico, and cause them to precipitate further north, over New Mexico, and thus, rather than from a diverted trade, they may have derived their unusual supply of moisture during the summer of 1854. On this subject I can but conjecture, and leave to future observation a discovery of the truth.

Enough appears, however, to show the importance of taking the location of the year in the decade, and even the character of the decade itself, into the account.

But whatever the remote cause of the difference in the seasons, the character of the seasons is directly influenced by the character of storms, or periodic changes. Sometimes the tropical storms are most numerous; at others the polar waves; and at others the irregular local storms, or general tendency to showers. The seasons when the polar waves are most prevalent, are the most regular, healthy, and productive. Those where the tropical tendency is greatest, are irregular; and so are those where the other classes predominate. These differences in the character of the storms, are but the varying forms in which magnetic action develops itself. I have said that there was a decided tendency to cirrus without cumulus, in mid-winter, and cumulus without cirro-stratus or stratus, in midsummer, and during the intermediate time an intermediate tendency. But there is a difference between spring and autumn. Dry westerly (not N. W.) gales prevail in March, and N. E. storms in April and May, but violent S. E. gales are not as common. On the other hand, the dry westerly gales of March are comparatively unknown in autumn, and the violent, tropical, south-easters are then common.

Snow-storms occur during the northern transit, not unfrequently in April and May; but they do not occur so near the acme of the northern transit on its return; nor until it approaches very near its southern limit. The quiet, warm, and genial air of April, is reproduced in the Indian summer of autumn, but they present widely different appearances. Those, and many other peculiarities of the seasons, deserve the attentive consideration of every one who would become familiar with the weather and its prognostics.

These irregularities in the character of the seasons have doubtless always existed, and always been the objects of popular observation. There are some very old proverbs which show this. I copy a few of the many, which may be found in Foster's collection. Mr. Graham Hutchison does not seem to think any of those ancient proverbs worthy of notice. But he misjudges. They are the result of popular observation, and many of them accord with the true philosophy of the weather.

Irregular seasons are unhealthy, and unreliable for productiveness. When the southern transit was late, or limited, and the autumn ran into winter, our ancestors feared the consequences in both particulars, and expressed their fears, and hopes also, in proverbs. Thus,

> "A green winter
> Makes a fat churchyard."

There is very great truth in this proverb. Again,

> "If the grass grows green in Janiveer,
> It will grow the worse for it all the year."

This is emphatically true, for the season which commences irregularly will be likely to continue to be irregular in other respects.

Another of the same tenor:

> "If Janiveer Calends be summerly gay,
> It will be winterly weather till Calends of May."

Janiveer is an alteration of the French name for January, and the proverb is very old.

So March should be normally dry and windy.

This, too, they understood, and hence the strong proverb:

> "A bushel of March *dust*
> Is worth a king's ransom."

And another:

> "March hack ham,
> Come in like a lion, go out like a lamb."

So April and May should be cool and moist. It is their normal condition in regular, healthy, and productive seasons. The grass and grain require such conditions; and the spring rains are needed to supply the excessive summer evaporation. This, too, they well understood. And hence the proverbs:

> "A cold April the barn will fill."

> "A cool May, and a windy,
> Makes a full barn and a findy."

And—

> "April and May are the keys of the year."

This was not very favorable, to be sure, for corn; but their consolation was found, as we find it, in the truth of another proverb:

> "Look at your corn in May, and you'll come sorrowing away;
> Look again in June, and you'll come singing in another tune."

This difference in the character of the seasons occasioned the adoption of a great variety of "Almanac days;" and they are still very much regarded. Candlemas-day (2d of February) was one of them.

Says Hone, in his "Every-Day Book":

"Bishop Hall, in a sermon, on Candlemas-day, remarks, that 'it has been (I say not how true) an old note, that hath been wont to be set on this day, that if it be clear and sunshiny, it portends hard weather to come; if cloudy and lowering, a mild and gentle season ensuing.'"

To the same effect is one of Ray's proverbs:

> "The hind had as lief see

> His wife on her bier,
> As that Candlemas-day
> Should be pleasant and clear."

St. Paul's day, or the 25th of January, was another great "Almanac day," and so the verse:

> "If Saint Paul's day be fair and clear,
> It does betide a happy year;
> But if it chance to snow or rain,
> Then will be dear all kinds of grain.
> If clouds or mists do dark the sky,
> Great store of birds and beasts shall die;
> And if the winds do fly aloft,
> Then war shall vex the kingdom oft."

St. Swithin's day was another of these "Almanac days." Gay said truly,

> "Let no such vulgar tales debase thy mind;
> Nor Paul, nor Swithin, rule the clouds or wind."

Yet *"Almanac days"* are still in vogue to a considerable extent—such as the *three first days* of the year, old style—the first three of the season—the last of the season—different days of the month—of the lunation, etc., etc. And some still look to the breastbone of a goose, in the fall, to judge, by its whiteness, whether there is to be much snow during the Winter, etc.

These *Almanac days should all be abandoned*; they have no foundation in philosophy or truth. There is one proverb, however, in relation to Candlemas-day, which the "oldest inhabitant" will remember, and which it may be well to retain. It has a practical application for the farmer, and in relation to the length of the winter:

> "Just half of your wood and half of your hay
> Should be remaining on Candlemas-day."

The months, too, have a character which must be remembered and regarded.

January is the coldest month of the year, in most localities. The atmospheric machinery reaches its extreme southern transit, for the season, during the month—usually about the middle. It remains stationary a while—usually till after the 10th of February. One or more thaws, resulting from tropical storms, occur during the month, in normal winters, but they are of brief duration. Boreas follows close upon the retreating storm with his icy breath. There is a remarkable uniformity in the progress of the depression of temperature, to the extreme attained in this month, over the entire hemisphere. It differs in degree according to latitude and magnetic intensity; but it progresses to that degree, whatever it may be, with as great uniformity in a southern as northern latitude. The table, copied from Dr. Forrey, discloses the fact, and so does the following one, taken from Mr. Blodget's valuable paper, published in the Patent Office Report for 1853:

TABLE SHOWING THE MEAN TEMPERATURE FOR EACH MONTH AT SEVERAL PLACES, VIZ.:

	Lat.	Jan.	Feb.	Mar.	Apr.	May.	Jun.	July.	Aug.	Sept.	Oct.	Nov.	Dec.
Quebec, Canada E.	46° 49'	9.9	12.8	24.4	38.7	52.9	63.7	66.8	65.5	56.2	44.1	31.5	17.3
New York, N. Y.	40° 42'	30.2	30.8	38.5	49.1	59.6	69.1	74.9	73.3	65.9	54.3	43.5	33.9
Albany, N. Y.	42° 39'	24.5	24.3	34.8	47.7	59.8	68.0	72.2	70.3	61.4	49.2	39.4	28.3
Rochester, N. Y.	42° 45'	26.1	25.8	33.0	45.8	56.2	64.5	69.7	67.8	60.1	47.7	38.2	28.8
Baltimore, Md.	39° 17'	33.1	34.3	42.4	53.0	63.2	71.6	76.6	74.5	67.7	55.8	45.0	37.8
Savannah, Ga.	32° 05'	52.6	54.7	60.0	68.4	74.8	79.4	81.3	80.6	76.9	67.2	58.3	52.2
Key West, Fla.	24° 33'	70.0	70.7	73.8	76.3	80.2	82.1	83.3	83.5	82.5	79.1	75.6	72.8
Mobile, Ala.	30° 40'	51.3	53.7	59.4	67.1	74.1	77.8	79.8	79.4	76.1	65.7	57.0	52.8
New Orleans, La.	30° 00'	54.8	54.5	61.5	67.6	74.0	78.6	80.4	79.6	77.1	69.1	57.5	56.2
Marietta, Ohio	39° 25'	32.2	34.1	42.6	53.0	61.8	69.2	72.7	70.9	63.5	51.8	42.6	34.7
San Antonio, Tex.	29° 25'	52.7	57.9	65.5	69.7	76.4	80.5	82.3	83.3	79.9	72.2	62.2	52.1
San Francisco, Cal.	37° 48'	50.1	51.0	53.8	57.7	55.9	58.8	57.9	62.2	61.6	61.9	56.2	50.0

Snows during this month are much heavier, and more frequent, in some localities than others. The reasons why this is so have been stated. The mountainous portions of the country receive the heaviest falls. They affect condensation somewhat, and according to their elevation. They intercept the flakes before they melt, and retain them longer without change. The thaws, or tropical storms, also sometimes have a current of cold air, with snow setting under them on their northern and north-western border. Such was the case with that investigated by Professor Loomis. January is without other marked peculiarities. It shows, of course, those extremes of temperature found, to a greater or less degree, in all the months, and differs, as the others differ, in different seasons. Normally, in temperate latitudes, it is a healthy month. The digestive organs have recovered from that tendency to bilious diseases which characterizes the summer extreme northern transit, and the tendency to diseases of the respiratory organs, which characterizes the southern extreme and the commencement of its return, is not often developed till February. February, in its normal condition until after the 10th, and about the middle, is much like January. Often the first ten days of February are the coldest of the season. The average of the month is a trifle higher, in most localities, as the tables show. This results from the increasing warmth of the latter part of the month. There are localities, however, where the entire month is as cold as January. Such (as will appear from Blodget's table) are Albany and Rochester, in the State of New York, and New Orleans, in Louisiana. At most places the difference is slight, either way. South of the latitude of 40° heavy snows are more likely to occur in the last half of January and first half of February than earlier. About the middle of the month we may expect thaws of more permanence in normal seasons. They are followed, as in January, by N. W. wind and cold weather, but it is not usually as severe. Many years since, an observing old man said to me, *"Winter's back breaks about the middle of February."* And I have observed that there is usually a yielding of the extreme weather about that period. Here, again, it is interesting and instructive to look at the tables, and see how regularly and uniformly the temperature rises in all latitudes, at the same time; as early and as rapidly at Quebec as at New Orleans or San Antonio; and subsequently rises with greatest rapidity where the descent was greatest. The elevation of temperature does not progress northwardly, a wave of heat accompanying the sun, but is a magneto-electric change, commencing about the same time over the whole country, and indeed over the hemisphere.

March is a peculiar month—the month of what is termed, and aptly termed, "unsettled weather." It, may "come in like a lion," or be variable at the outset. The northern transit is fairly started, and is progressing rapidly, and there is great magnetic irritability. A reference to the table of Dr. Lamont will show that the declination has increased with great rapidity. Normally, the early part is like the latter part of February, and the latter part approaches the milder but still changeable weather of April. Its distinguishing feature is violent westerly wind. Not the regular N. W. only—although that is prevalent—but a peculiar westerly wind, ranging from W. by N. to N. W. by W., often blowing with hurricane violence. This wind was alluded to on page 130. With the change and active transit to the north, in February and in March, comes the tendency to diseases of the respiratory organs—pneumonias and lung fevers—and this is the most dangerous period of the year for aged people.

April is a milder and more agreeable month. During some period of it, in normal seasons, and at other times in March, there is a warm, quiet, genial, "lamb-"like *spell*, exceedingly favorable for oat seeding. When it comes, advantage should be taken of it, for long heavy N. E. storms are liable to occur, and frequently with snow. On the latitude of 41° heavy snow-storms are not uncommon in April. Within the last fifteen years two such have occurred after the 10th of the month. April, as we have seen, should be cool and moist. If dry, the early crops are endangered by a spring drouth; if very wet, there is danger of an extreme northern transit, and an early summer drouth. It is emphatically true that

"April and May are the keys of the year."

Its distinguishing peculiar feature is the gentle, *warm, trade* rains—*"April showers"*—which, in the absence of great magnetic irritability, that current drops upon us. There is great *mean* magnetic activity, but it is not so *irregularly excessive* as in March.

May, in our climate, should be, and normally is, a wet month, and a cool one, considering the altitude of the sun. The atmospheric machinery which the sun moves is, however, ordinarily about six weeks behind it—the latter reaching the tropic the 20th of June, and the former its farthest northern extension about six weeks later. Hence it is not a cause for alarm if May be wet and cool. The great staples, wheat, grass, and oats, are benefited; and corn, according to the proverb, will not be seriously retarded. The movable belt of excessive magneto-electric action, with its tropical electric rains, so exciting to vegetation, and its periods or terms of excessive heat, is on its way north, and sure to arrive in season, and remain long enough to mature the corn. There have been but two seasons in this century when corn did not mature in the latitude of 41°. One during the cold decade, and the cold part of it, between 1815 and 1820; and the other, during the cold half of the fourth decade, between 1835 and 1840.

The distinguishing feature, if there be one, of May, is its long, and, for the season, cool storms. These have, in different localities, different names. In pastoral sections we hear of the *"sheep storms"*—those which effect the sheep severely when newly shorn—killing them or reducing them in flesh by their coldness and severity.

In relation to this too early shearing, there is an old English proverb, in "Forster's Collection," viz.:

> "Shear your sheep in May,
> And you will shear them all away."

So there are others called *"Quaker storms,"* which occur about the time when that estimable sect hold their yearly meeting. And there are other names given in different localities to these long spring storms. But they are all *mere coincidences—* equinoctial and all.

Notwithstanding the storms, however, the temperature rises at a mean. The declination is often as great as in mid-summer. The earth is growing warmer by the increase of magneto-electric action, whatever the state of the atmosphere. The yellow, sickly blade of corn is extending its roots and preparing to *"jump"* when the atmosphere becomes hot, as it is sure to do, when the machinery attains a sufficient altitude, how backward soever it may seem to be. The farmer need not mourn over its backwardness, unless the season is a very extraordinary one, like those of 1816 and 1836. The storms ensure his hay, wheat, and oat crops; the warming earth is at work with the roots of his corn, and is filling with water, and preparing for the hot and rapidly-evaporating suns of mid-summer. The earth would grow warmer if every day was cloudy.

By the middle of June the atmospheric machinery approaches its northern acme, the summer sets in, and not unfrequently, as extremely hot days occur during the latter part of the month, as at any period of the summer. But the heat is not so continuous, or great, at a mean.

From the middle of June to the latter part of August is summer in our climate, and during that period from one to three or four terms of extreme heat occur, continuing from one to five or six days, and possibly more, terminating finally in a belt of showers overlaid with more or less cirro-stratus condensation in the trade, and controlled by the S. E. polar wave of magnetism, and followed by a cool but gentle northerly wind. During these "heated terms," a general showery disposition sometimes, though rarely, appears, with isolated showers, which bring no mitigation of the heat. Not until a southern extension of them appears, followed by a N. W. air, does the term change, so far as I have observed.

By the 20th of August, in the latitude of 42°, an evident change of transit is observable, by one who watches closely, although the range of the thermometer in the day-time may not disclose it. A greater tendency to cirrus-formation is visible. The nights grow cooler in proportion to the days. The swallows are departing, or have departed; the blackbirds, too, and the boblinks, with their winter jackets on, *their plumage all changed to the same colors*, are flocking for the same purpose, and hurrying away. The pigeons begin to appear in flocks from the north, and the first of the blue-winged teal and black duck are seen straggling down the rivers. At this season, and nearly coincident with the change, the peculiar annual catarrhs return. These are colds (so called) which at some period of the person's life were taken about or soon after the period of change, and have returned every year, at, or near the same period. They soon become *habitual*, and no care or precaution will prevent them. I know one gentleman who has had this annual cold in August for twenty-seven years, with entire regularity; and another who has had it nineteen years; and many others for shorter periods. I never knew one which had recurred for two or three years that could be afterward prevented, or broken up. *Very instructive are these annual catarrhs* to those who think health worth preserving, and in relation to the change of transit.

The change is felt over the entire hemisphere. Between the 20th of August and the 10th of September hurricanes originate in the tropics and pursue their curving and recurving way up over us; or long "north-easters" commence in the interior and pass off to E. N. E. on to the Atlantic, followed now in a more marked degree by the peculiar N. W. wind, so common over the entire Continent in autumn and winter.

By the 10th of September the pigeons may be seen in flocks in the morning, and just prior to the setting in of a brisk N. W. wind, hurrying away southward with a sagacity that we scarcely appreciate, to avoid the anticipated rigors of winter, and to be followed soon by all the migratory feathered tribes that remain.

The nights grow cooler, although the sun shines hot in the day-time, and woe to the person, unless with an iron constitution, who disregards the change, and exposes himself or herself without additional protection, to its influence. Nature has taken care of those who depend upon her, or upon instinct, for protection. The feathers of birds and water-fowl are full; the hair and the fur are grown. Beasts and birds have been preparing for the change, and are ready when it begins. They know that the earth is changing. The shifting machinery is fast carrying south that excess of negative electricity which has so much to do with giving it its summer heat. They feel its absence, even during the day, and the contrast between that and the positively electrified northern atmosphere, which now follows every retreating wave of condensation.

The musk-rat builds, of long grass and weeds, his floating nest in the pond, that he may have a place to retire to, when the rain fills it up and drives him from his burrow in its banks.

But man, with all his intellect, is too heedless of the change. Additional clothing is now as necessary to him as to animals, but it is burdensome to him in the day time, and therefore he will not wear it, how much soever it would add to his comfort and safety during the night. He stands with his thin summer soles upon the changed ground, or sits in a current, or in the night air, less protected than the animals, and dysentery or fever sends him to his long home. He has *intelligence*, but he lacks *instinct*. He has time for the changes of dress which fashion may require, but none for those which atmospherical changes demand. *Fashion* has attention in *advance*; *death* none till *at the door.*

Now the southern line of the extra-tropical belt of rains descends upon those who, living between the areas of magnetic intensity, have a dry season; and the focus of precipitation in that belt descends every where. *"Winter no come till swamps full,"* the Indians told our fathers, and there is truth in the remark; although like other general truths respecting the weather, it is not always so in our climate. Rains fall during the autumnal months, as during the spring months, and while the transit of the machinery is active and the evaporation is less. And the magnetic comparative rest, and the seed time and equable "spell" of April is reproduced in the Indian summer of autumn.

The machinery gradually and irresistibly descends, and with an excess of polar positive electricity, comes snow; Boreas controls, and winter sets in, reaching its maximum of cold in January again.

Remembering, then, the differences in the normal conditions of the seasons and months, and the different characters that the winds, and storms, and clouds, and other phenomena bear in them respectively, let us now look at the signs of foul or fair weather not herein before fully stated, upon which practical reliance may be placed.

In the first place, we must look to the forming condensation. There are many days when the atmosphere is without visible clouds, but few when it is entirely without condensation. Such days are seen during the dry season in the trade-wind region; and with us, in mid-summer drouths, which partake of this tropical character; and when, at any season, but particularly in winter, the N. W. wind in large volume has elevated the trade very high. Condensation is not necessarily in form of visible cloud. It may be of that smoky character which sometimes attends mid-summer drouths, giving the sun a blood-red appearance; or it may be like that change from deep azure to a "lighter hue," obscuring the vision, which Humboldt describes as preceding the arrival of the inter-tropical belt of rains. Gay-Lussac, and other aeronauts, have seen a thin cloud stratum at the height of 20,000 to 30,000 feet, not visible at the earth, although some degree of mistiness and obscurity were observed. At that elevation the clouds are thin, and always white and positive. Some degree of turbidness is frequent; it may occur, as we have stated, with N. W. wind, but, if it does, the wind soon changes round to the southward.

This turbidness or mistiness, where it exists, and indicates rain, does not disappear toward night, as it should do if but the daily cloudiness which results from ordinary diurnal magnetic activity, but becomes more obvious at nightfall; and, when hardly visible at mid-day, or during the afternoon, may then be observed, obscuring in a degree, the sun's rays; and, later in the evening, forming a circle round the moon. Thus Jenner—

> "Last night the sun went *pale to* bed,
> The moon in *halos* hid her head."

And so, too, Virgil—

> "The sun, too, rising, and at that still hour,
> When sinks his tranquil beauty in the main,
> Will give thee tokens; certain tokens all,
> Both those that morning brings, and balmy eve.
>
> *******
>
> When Sol departs, his mighty day-task done,
> How varied hues oft wander on his brow.
>
> *******
>
> If the ruddy blaze
> Be *dimm'd* with *spots*, then all will wildly rage
> With squalls and driving showers: on that fell night
> None shall persuade me on the deep to urge
> My perilous course, or quit the sheltering pier.
> But if, when day returns, or when retires,
> *Bright* is the orb, then fear no coming rain:
> Clear northern airs will fan the quiv'ring grove.
> Lastly, the sun will teach th' observant eye
> What vesper's hour shall bring; what clearing wind
> Shall waft the clouds slow floating—what the South
> Broods in his humid breast. Who dare belie
> The constant sun?"

More frequently this kind of condensation is sufficiently dense at night-fall to take shape, and show a bank when the sun shines horizontally through a mass of it. I am now speaking of *storm* condensation, or that which indicates the approach of a storm. Thunder clouds at nightfall, dark, dense, and isolated, are, of course, to be distinguished. Those, every one understands to indicate a shower, and immediate succeeding fair weather.

The halos do not, in cases of incipient storm condensation, always appear. The moon may not be present: though, in her absence, I have seen them in the light of the primary planets; or she may be in the eastern portion of the heavens. When this is so, and the condensation forms slowly, there may be less appearance of it, after the sun disappears, than before, although a storm is approaching, and sure to be on by the middle of next day, and perhaps with great violence. When the failure of the light no longer reveals the denser condensation in the west, the stars may

shine, as did the sun, dimly but visibly, through the partial and invisible condensation; and one who did not notice the bank in the west, at nightfall and before
dark, may be deceived by the seeming clearness of the evening. Thus Virgil—

> "Mark, with attentive eye, the rapid sun—
> The varying moon that rolls its monthly round;
> So shalt thou count, not vainly, on the morn;
> *So the bland aspect of the tranquil night*
> *Will ne'er beguile thee with insidious calm."*

All early condensation and indications derived from it, must be looked for in
the west. From that quarter all storms come. These indications at nightfall are of
a varied character. They may consist of primary condensation in the trade, or of
secondary condensation, scud running north toward a storm, the condensation of
which has not yet visibly reached us, but which will extend south and pass over
us. It may be a heavy bank, or consist of narrow cirrus bands. Cirro-stratus cloud
banks, in the S. W., in the fall and winter, of a foggy and uniform character, are
indicative of snow. The body of the storm will pass south of us, and a portion over
us, the wind be north of east, and the snow will not be likely to turn to rain before
it reaches the earth, by reason of a southern middle current.

Banks in the N. W. indicate rain at all seasons. The storm is north of us, working southerly, and such storms rain on the southern border—in winter even—
because they have the wind on that border from south of east. It may, indeed,
snow, but if so, probably in large flakes, soon turning to rain. There are other
appearances at nightfall which deserve consideration. A red sun, with smoky air,
is indicative of continued dry weather, a frequent appearance in dry terms, lasting
three or four days, at least, from the commencement. So is a red appearance of the
sky, when there are no clouds, indicative of a fair day following. On this subject we
have an allusion to the weather, by our Saviour while on earth, which, like all such
allusions found in the Bible, is of remarkable philosophical accuracy. It is found in
Matthew, chapter xvi., verses 2 and 3: "He answered and said unto them, When it
is evening ye say, It will be fair weather, for the sky is red. And in the morning, It
will be foul weather to-day, for the sky is red and lowering. O, ye hypocrites, ye
can discern the face of the sky," etc.

Another allusion to the weather, though not applicable to this point, I will
refer to in passing. It is found in Luke, chapter xii., verses 54 and 55: "And he said
also to the people, When ye see a cloud rise out of the west straightway ye say,
There cometh a shower; and so it is. And when ye see the south wind blow, ye say,
There will be heat; and it cometh to pass."

This is all very true, and might have been cited to show the universality of the
phenomena. But to return.

We have an old English proverb alluding to the same phenomena, of great
value and truth, viz.:

> "An evening red and a morning gray
> Are sure signs of a fair day;
> Be the evening gray and the morning red,

Put on your hat or you'll wet your head."

The sky is red if there be no condensation at the west to obscure the rays of the sun; if there be, it is gray, or there is a bank or cloud, and it is obscured. So if there be no condensation over, or to the east of us, in the morning, to reflect the rays of the sun, the sky is gray; if there be such condensation, the sun is reflected from it, and the sky is red. Such morning condensation is indicative of foul weather. It is, as we have said, the eastern edge of an approaching storm, on, or under which, the sun shines and illumines it. Thus, at night, it shines through a portion at the west, which is situate between the sun and us, making the sky gray: but shines on, or under, a portion in the morning, east of us, but not far enough east to obscure the horizon, and the rays of the rising sun are reflected from it. In either case the red or gray appearance results from the relative situation of the sun and the eastern edge of an approaching storm.

The following couplet of Darwin is an apt description of the morning appearance:

> "In fiery red the sun doth rise,
> Then wades through clouds to mount the skies."

The sun is often reflected in vivid colors, from the under surface of clouds, at sunset. This is an indication of fair weather. It is evident the sun shines through a *clear atmosphere beyond the cloud*, or his rays would not reach and illume the lower surface of the cirro-stratus with such distinctness. He *"sets clear,"* as is said; the clouds are passing off, and there are none beyond. It is this appearance, in different forms, when there happen to be patches of broken, melting cirro-stratus above the horizon, which makes the beautiful sunsets that attract attention. So the sun is reflected, in beautiful colors sometimes, from the cumulus clouds which have passed over to the east. The most beautiful and variegated I have ever seen, were reflected from that imperfect cumulus condensation which takes place occasionally during long drouths—doubtless resembling that which is seen over Peru, hereinbefore alluded to, as described by Stewart.

It is not, then, the presence of cloud condensation at the west, at nightfall, which alone indicates foul weather; but such condensation, whatever its form, as evinces that it is not the *dissolving* cloud of the day, but the eastern, approaching portion of a *still denser portion beyond, through, or under which, the sun can not shine clearly, but which wholly or partially obscures it. Remembering this philosophy of the matter*, the observer will soon be able to detect the various forms of condensation which originate or exhibit themselves at nightfall, and whether they indicate an approaching storm or not, without a more explicit specification of them. It is an important hour for observation; "Let not the sun go down" without attention.

When the condensation is obvious, but thin, at nightfall, it may not, as I have said, be discernible in the evening. But there are methods by which the incipient storm condensation may be detected. The number of the stars visible, and the *distinctness* with which they may be seen, indicate the absence or presence of condensation and its density. Virgil, alluding to the indications of fair weather, says:

> "*Brightly* the stars shine forth; Cynthia no more

Glimmers obnoxious to her brother's rays;
Nor fleecy clouds float lightly through the sky."

The brightness of the stars and the clear appearance of the moon show the absence of condensation and the *dissolution* of the fleecy clouds at the close of the day is, as we have seen, always a fair-weather indication.

There is much true philosophy in the allusions of Virgil to the moon. Thus—

"When Luna first her scatter'd fires recalls,
If with *blunt horns* she holds the *dusky* air,
Seamen and swains predict th' abundant shower."

The horns, or angles of the moon will, of course, appear distinct and sharp or indistinct and blunt, in proportion to the amount of condensation in the atmosphere which impedes the passage of the light. For the same reason, when the moon is new, her entire disk is visible when the atmosphere is very clear, by reason, as is supposed, of light reflected from the earth to the moon and back to us. This double reflection can only take place when the atmosphere is very clear. Hence, Virgil alludes to it, and correctly, as an indication of continued fair weather:

"If (mark the ominous hour!)
The clear fourth night her lucid disk define,
That day, and all that thence successive spring,
E'en to the finished month, are calm and dry."

Probably Virgil alluded to a month of the summer trade-wind drouth which reaches up on Southern Italy. But that appearance of the moon is occasionally seen here, and the indication is, in degree, philosophically true.

It is somewhat more difficult to determine what will be the result of the condensation seen at the west in the morning, and which is not so far east, or of such a character, as to reflect the rays of the sun; for, although always suspicious, it is sometimes of a foggy character, and disappears between eight and nine o'clock. If it increases in density after ten o'clock, or is of a dense cirro-stratus character, rain may generally be expected. If of a decided *cirro-cumulus* character, it is certain to disappear. Cirro-cumulus is seen in small patches, with small, distinct, and rounded masses, in summer, in the morning, and sometime, during the day, after high fog has disappeared, and at other times, and is always, when of that *distinct* character, a fair weather indication. I have seen it thus when the wind was blowing from the N. E., and the scud running toward a storm passing near, but to the south of us, when those who relied upon the existence of the wind and scud as evidences that we were to have the desired rain, were deceived. Thus, the couplet from an old almanac:

"If *woolly fleeces* strew the heavenly way,
Be sure no rain disturb the summer day."

When this morning condensation is not high fog, and is dense and passing east with a wavy appearance, it is very certain to rain. Jenner says:

"The boding shepherd heaves a sigh,

For see, a rainbow spans the sky."

An old almanac had the following verse:

"A rainbow in the morning
Is the shepherd's warning;
A rainbow at night
Is the shepherd's delight."

So the proverb was originally made; but as our ancestors were not shepherds, and had a horror of ocean storms, it was commonly quoted, in this country, in the following form:

"A rainbow in the morning,
The sailors take warning," etc.

Rainbows are not reflected from *clouds*, but falling rain, and a morning rainbow at the west is, of course, evidence that it is *actually raining there*, and will, in all probability, pass over us. "Thunder in the morning, rain before night," is a common saying, and a true one. There is a belt of showers, or showery period approaching, of unusual intensity—for thunder showers in the morning are rare. The afternoon is their most common period, and they are very apt to appear then, when the morning is showery.

Of the different forms of cirrus and cirro-stratus, which appear during the day, and indicate approaching storms, or of cumulus indicative of showers, it is difficult to give an intelligible description without very many illustrations. I have many daguerreotype views, taken at different seasons of the year, and at a time when different forms of cirrus and cirro-stratus condensation, indicative of storms, exhibited themselves. They differ, as I have said, and it must be remembered, very much at *different seasons* of the year, and in *different years*, and their delicate shades are taken with difficulty by the artist, and reproduced with difficulty, and only at considerable expense, by the engraver; and I have omitted them. The time will come when a knowledge of their language will be sought for and read—when the "countenance of the sky" will be an object of intelligent interest to all whose business may be affected by the weather, or who love to learn of nature. But it is not yet. This is the age of theory and speculation. The time of actual, practical, connected observation and prognostication, which may justify expensive illustration, is yet to arrive.

The reader will find in the general plates representations of several kinds of cirri. They are delicate, always white, more or less fibrous, and form in the upper part of the trade or the adjoining atmosphere above it. Their character and elevation should be studied, and the observer should be careful to distinguish which is the most elevated. Not unfrequently it may seem, to a hasty observer, that the cirrus is below the cirro-stratus or forming stratus. But the genuine cirrus never is. It forms near, and above, the point of congelation, and is often composed of crystals of ice or snow. If they fall, they melt and evaporate, when there is no storm, before reaching the earth. Aeronauts have met with them and their crystals when there was no fall of moisture at the surface of the earth; and the angles of reflection exhibited by halos and other optical phenomena which form in them, enable us to detect their crystallization and the form of it.

They are produced by electric changes which condense the vapor, and the coldness of the air at that elevation freezes it at the *instant of its condensation*.

Congelation is crystallization, and all crystallization is electric, or magneto-electric. The snow-flakes differ in form and size according to the suddenness of the condensation, the amount of moisture condensed, the polarity of the strata through which they pass, and their consequent attraction and adhesion to each other.

The connection of electricity with these formations of cirri has frequently been admitted, and it is perfectly obvious that the long fibrous bands, shooting from horizon to horizon, could not be formed by commingling of currents any more than the perfectly isolated, distinct, enlarging-outward cumulus hail-storm, could be so formed. Cirri form at the line of meeting, between the trade and the upper atmosphere, and in one or the other, or both, very much according to the season, and the suddenness with which storms are produced. These often *induce* a layer of cirro-stratus or stratus at the lower line of the counter-trade, and in the surface-atmosphere, which precipitates; and this operation is clearly discernible, and very frequently, before gentle rains. Condensation in the whole body of the trade is usually in the form of turbidness or mistiness, a bank or incipient stratus, without cirri.

It seems matter of astonishment that water should float so far condensed, in strata where the air is so much lighter, without being precipitated. But electric attraction and repulsion between the different strata and the vesicles, explain it.

In mid-winter, the cirrus forms are prevalent and most distinct. After severe cold weather, when a storm approaches, the cirri form in long, narrow threads, parallel to each other, extending from about W. S. W. to E. N. E., gradually thickening and forming, or inducing, cirro-stratus and stratus, and dropping snow. This form is called the *linear*-cirrus. The tufted, and other fibrous forms, are seen in patches also, in great distinctness, during these mid-winter days, when the wind gets around to the southward, and the weather is pleasant. Such days are called *"weather-breeders,"* and their *offspring* the patches of cirrus, which are to extend and compose, or induce the storm, and indeed are an advance part of it, are then never absent. A clear, moderate day, in a normal winter, with wind from any southern point, however light, between the 1st of January and the middle of February, without these patches of cirrus, is very uncommon. Watch and see whether they tend to cirro-stratus, or whether the wind gets around to the N. W. at nightfall, and they disappear. If the former, a storm may be expected; if the latter, fair weather.

Thus there are three peculiarities attending the forming cirrus of mid-winter (1st of January to 10th of February): long, fibrous, parallel bands in the morning (linear cirrus), gradually coalescing as the day advances, after severe cold; the comoid, curled, or tufted cirrus, in curling bunches, called *"mares'-tails,"* and the *transverse*, when the fibers are in bands or threads, which are not parallel, but cross each other at angles, more or less acute. The two former varieties are represented on Figure 5, page 26, indicated by one bird, but the last form is a very prevalent one in our atmosphere.

Various names have been given to different forms of *cirro-stratus*. Those represented in Figure 5, page 26, are the *"cymoid"* on the right, the *"mottled"* on the left, below the cirro-cumulus; and the *"linear"* below that. The form known as the *"mackerel sky"* is not represented there. It consists of regular forms, resembling the *waves* on the surface of the water when the wind blows a gentle breeze. But the *wavy* form, and of all sizes, is very frequently assumed by cirro-stratus, which is rapidly condensing, and turning to stratus. In the "mackerel sky," strictly so called, the waves are small, parallel, nearly distinct and equi-distant, and resembling the appearance of a school of mackerel, swimming in the same direction, one above another. All *wavy* forms of cirro-stratus indicate a disposition to increased condensation and rain. When the waves are very large and dense, and cross obliquely, or unite at one end, rain is very certain to fall soon, if the line of progress of the condensation is over the observer, and the clouds are seen in the western or N. W. quarter of the sky.

But there are few forms which are not occasionally seen when no rain or snow falls. The intensity of the electric action which produces them may not be sufficient to effect precipitation, or they may be the attendant, attenuated *lateral* condensation, which frequently "thins out" a considerable distance from the dense, precipitating portions of the storm.

If that denser portion is north of us, the probabilities of rain are greater, for there is always a probability that the storm may be of the character which is extended south, by a polar wave. The observer must watch the formation of cirri, and the different forms of cirro-stratus and stratus, and become familiar with their appearance. It is not a difficult task. With the aid of a few general directions he will soon be familiar with them:

1. Get a correct idea of the different characters of the primary clouds. The true fibrous *cirrus*—the different forms of *cirro-stratus*—the smooth, uniform *stratus*—the *cirro-cumulus*, which is nothing but a cirro-stratus, separated into *distinct masses* by the repulsion of static electricity—and the *cumulus*, too distinct ever to be mistaken. There is no difficulty, except with the varied forms of cirro-stratus. It is useless to attempt to give, or the observer to rely on, names for these numerous forms, without as numerous illustrations. Those in use are rarely applied correctly. I have never met with ten persons who applied even the term "mackerel sky" to the same precise form of cirro-stratus. In relation to all of them it is to be observed that polar belts of condensation, and local appearances of considerable extent, are often too feeble in action to precipitate, even when the mackerel form is present; and all may be the lateral attendants of passing storms. Therefore,

2. Satisfy yourself whether the cirrus or cirro-stratus increases in density and tends to the formation, or induction, of stratus; and whether it is isolated, or an extension of the condensation of a storm, and if the latter, *where that storm is*. The time will come when an intelligent use of the telegraph will do this for you.

3. Look also to the character of the wind, if there be any. On this subject I have perhaps said all that is necessary in the preceding pages. Next to condensation, the

direction and character of the wind is the most valuable prognostic. Indeed it often tells us that a storm is approaching, and the quarter from which it will come, and its character, before the condensation is visible.

4. See if there is any *secondary* condensation or scud. These are sometimes seen running toward a storm, when there are not distinct clouds visible in the western horizon, at nightfall, or in the evening, as in the instance stated in the introduction, and sometimes from the north-east, as in cases heretofore so often stated. But the easterly scud do not often form in winter, until after the cirrus has passed into the form of cirro-stratus, or has induced the latter forms in the inferior portion of the trade, or the surface atmosphere.

The inductive effect of the primary condensation, therefore, is not always, and especially in winter, sufficient to create the easterly current and scud, and it is often the case that the easterly wind is not felt, or the scud seen, in snow-storms, until the snow has begun to fall, and the first snow will fall with a S. W. air, as I have heretofore stated. But when the condensation has so far advanced toward stratus that the easterly wind and scud are obvious, there is little or no doubt that rain or snow will fall speedily. The occasional occurrence of easterly wind and scud, without rain, however—dry north-easters, as I have termed them—in connection with storms passing south of us, or condensation too feeble to precipitate, should be remembered. The long, dry, north-easterly winds of spring have been attributed to the icebergs, but they are overlaid by feeble stratus or cirro-stratus condensation, or are the result of attraction, by a more southern precipitation. The observer must be careful to distinguish between the various forms of N. W. scud and cirro-stratus, which they sometimes resemble. This he may do *from the direction in which they move*. Cirro-stratus always moves from some point between S. S. W. and W. S. W. to some point between N. N. E. and E. N. E. The various forms of N. W. scud move to the S. E. The March, foggy scud, from between W. and N. W., rarely have any cirro-stratus above them, but rather a peculiar turbid condensation.

The character of the primary condensation, the direction and force of the wind, and the direction of the secondary condensation or scud, must be the main reliance of the observer. But I must reiterate that they all differ in different kinds of storms, in different seasons of the same year, and the same seasons of different years; and the observer must be careful to make due allowance for those differences.

There are, however, divers other secondary signs, which, although not alone to be relied upon, will aid the observer, if carefully studied, when the character of the clouds, and the pressure of easterly or southerly wind and scud, are not decisive. Of these, a large class are electrical.

The smoke descends the adjoining chimney-flues, or outside of the chimney, toward the ground.

Thus, Darwin, as quoted by Hone:

> "The smoke from chimneys right ascends,
> Then, *spreading*, back to earth it bends."

Smoke is electrified *positively*, by the act of combustion; the earth and the adjacent atmosphere, when storms are gathering or approaching, is *negative*. Hence the smoke spreads, and is attracted downward by an opposite electricity. On the other hand, it is interesting to see, at other times, and when the difference in temperature is not material, but the whole atmosphere is positive, with what rapidity and compactness the smoke will ascend in a *straight and elevated column* from the chimney, repelled by a similar electricity. I am aware it is generally supposed the smoke descends because the *air is lighter*. But it is a mistake. I have seen it descend when the barometer was at 30°.60, or .60 above the mean.

There is, too, a draught downward in chimneys, in such cases when there is no smoke or fire in any of its flues. Thus Jenner says: "The soot falls down;" whether he meant by this that there was an actual fall of soot other than what is occasioned by the rain falling in through the chimney top, and disturbing the soot, as sometimes happens, I do not know. It occurs rarely, and is of very little practical importance. But every housewife knows that chimneys, which have been used in winter, and are full of soot, *smell* before storms. The odor results from a downward draught and the dampness of the air. So the smoke from one flue will descend another, into some unused room, on such occasions. Another class of these electrical signs are felt by those who are suffering from chronic diseases, which have affected the nerves and made them sensitive. Thus Jenner:

"Old Betty's joints are on the rack."

And Hone adds:

"Her corns with shooting pains torment her,
And to her bed untimely send her."

But Old Betty's rheumatism or corns are not alone in this. Those whose bones have been broken feel it. All invalids feel it. And, indeed, all observing healthy persons may, and do, although all are not distinctly conscious of it. It is common for such to say, I feel sleepy, or I feel dull, or, It *feels* like snow, or *feels* like rain, and thus from their own feelings to be able to predict, not only falling weather, but its *character*, whether snow or rain, at a time when either may occur consistently with appearances.

This change is a change from the positive electricity which is so congenial to the active—"bracing" is the usual term—to negative and damp—for this change is accompanied by condensation, as I believe all changes from positive to negative are. Certain it is, if the atmosphere is highly charged with negative electricity, condensation takes place; if with positive, evaporation. Perhaps it is a change of the associated electricity which accompanies magnetism, and not of the free atmospheric electricity alone. Hence another phenomenon alluded to by Jenner:

"The walls are damp, the ditches smell."

There are localities where this dampness is very obvious. The celebrated William Cobbett, many years since, when a farmer on Long Island, observed and published the fact that the stones grew damp before a storm. I know of flagging stones that usually grow damp two or three hours before rain, especially in spring and fall, and every step taken upon them is made visible by a corresponding increase of condensation.

The reverse of this takes place just before the close of storms. Flagging stones, and walls under cover, will frequently become dry before the rain ceases. The negative electricity becomes less as the positive prevails, although the clouds above are still dropping rain.

In the comparatively moist, showery climate of England, these changes from positive to negative alternate rapidly between successive showers; but observations of electric phenomena, or of clouds, in that climate, are not, without qualification, safe guides for us.

So "the ditches smell," particularly in the evening before a rain, when the immediate surface-atmosphere is charged with negative electricity, and the *condensing moisture* prevents the diffusion of the odors. For the same reason the candle will not relight, and there is crackling in the ashes or lamp. Thus, again, Virgil:

> "Maidens that nightly toil the tangled fleece
> Divine the coming tempest; in the lamp
> *Crackles* the oil, the gathering wick grows dim."

Virgil did not live in our cold climate, and knew nothing of the crackling in the fire, or in the ashes or coals which remain after the wood is consumed. The lamp exhibits it on a smaller scale, and perhaps he had noticed it when in company with the maidens. But it is sometimes noticeable even in the lamp or candle with us. A small particle of moisture will produce it, in a marked degree, at any time.

In winter, when the air is highly positive and cold, the candle can be blown out, and by another puff of the breath relighted, with ease. But when the electricity before a storm becomes negative, and partial condensation takes place, this can not be done. This partial condensation before storms and showers shows itself upon vessels containing cold-water, in summer. It seems to be the received opinion, that the condensation is evidence of a greater *quantity* of moisture in the atmosphere. But this, too, is a mistake, and hence the little reliance to be placed on hygrometers.

This partial condensation is sometimes visible. When the sun shines clearly, at the east or west, through a *small opening* in the clouds, the condensing vapor is shown by the streaks of sunlight, just as the fine particles of dust are seen in a dark room, when a few rays of sunlight are admitted through a small aperture. This phenomenon is often observed, and it is said of it—"It's a going to rain; *the sun is drawing water.*"

Virgil alludes to this as seen in the east in the morning, thus:

> "But when beneath the dawn *red-fingered rays*
> Through the dense band of clouds *diverging* break,

****

> Ill does the leaf defend the mellowing grape;
> Leaps on the noisy roof the plenteous hail,
> Fearfully crackling."

It is well ascertained that storm-clouds of great intensity have polarity in the different portions, and that in the less intense magneto-electrical climate of England isolated showers are often of this character—the polarity existing in rings. Showers are doubtless thus found with us. Mr. Wise got into one of them; see his description (Theory and Practice of Aeronautics page 240).

I have, in another place, alluded to the upward attraction of the dust beneath the advance condensation of a shower. Jenner alludes to it in the following lines:

> "The whirling winds the *dust* obeys,
> And in the rapid eddy plays."

So Virgil:

> "Light chaff and leaflets, *flitting, fill the air,*
> And sportive feathers circle on the lake."

All these are electrical.

In England, where the action of such isolated clouds is less intense, the different electricities in different portions of the cloud, whose opposite and changing action produce all the phenomena, the condensation, the cold and congelation, the currents, etc., have been accurately ascertained. We can not get into the situation occupied by Mr. Wise. But every man may observe these *intestine motions* occasionally, in the advance condensation of an isolated thunder-shower, in front of, but near the smooth line of falling rain. They are more lateral than upward or downward, and are often exceedingly rapid in movement.

I have said that hail has often been found to fall from particular and well-defined portions of a cloud, and rain from the other portions, the hail being positive, and rain negative. An instance of very striking character may be found in Espy's Philosophy of Storms (Introduction, page xx.) Doubtless in all cases thunder-showers, which are isolated and distinct, have opposite electricity in different portions, to whose active agency all the phenomena are owing. And the return of electricity to the earth in the rain explains the greater fertilizing effect of the latter compared With all artificial watering. He was a true philosopher who attempted to stimulate vegetation by electricity.

Sounds may sometimes aid the observer in doubtful cases in foretelling the weather. The roar of the surf, or breaking of the waves on the shore, when great bodies of water are disturbed by a precedent storm-wind, often heard before the wind is perceived on the land, I have already alluded to. And thus Virgil:

> "When storms are brooding—in the *leeward gulf*
> Dash the swelled waves; the mighty mountains pour
> A harsh, dull murmur; far along the beach
> Rolls the deep rushing roar."

The moaning or whistling of the wind all have noticed. It is not uncommon to hear the expression, "The wind sounds like rain." Jenner says:

> "The *hollow* winds begin to blow."

And Virgil:

> "The *whispering* grove
> Betrays the gathering elemental strife."

This whispering is the motion of the leaves; and they are often stirred by a peculiar motion which is not that of wind. Sometimes every leaf upon a tree may be seen *vibrating* with an *upward and downward* motion, when there is not wind enough to stir a twig. This interesting phenomenon is electrical. Trees, and all vegetables, confessedly discharge electricity, and such discharges move the leaves, when very active.

With us, sounds can be heard more distinctly from the east or south, before storms, according to the character of the coming wind. Howard mentions an instance when he heard carriages five miles off. Steamboat paddles, rail-road cars, and other sounds, are often heard a great distance. The distance at which the now common steam-whistle is heard, and the direction, is not an unimportant auxiliary indication of the weather. Howard attributes these peculiar phenomena to the *"sounding board,"* made by the *stratum of cloud*; but sounds may be heard from the north-west, when there is no condensation, and the wind is from that quarter, and also from the east when it is not cloudy; and in a level country the village bells often tell the direction of the current of air just over our heads when we do not feel it at the surface. The wind is undoubtedly moving in a rapid, and perhaps invisible current, not far above us. If from the east or south, it betokens rain; if from the western quarter, fair weather.

The conduct of the different animals furnish a considerable portion of the signs alluded to by Virgil and Jenner, and are never unimportant auxiliary evidence of the approaching changes, whether from dry to wet, or wet to dry.

The observer will find, in the conduct of our birds and animals, especially those which are not domestic, ample evidence of the truth of the descriptions of Virgil. He denies the animals and birds foresight, but he does not seem to have observed that the swallow leaves for the south as soon as the *autumnal* change begins to be felt, and in August; nor the evident sagacity of other *migratory* birds. They do not act from the *"varying impulse"* produced by an actual state of things, but a knowledge or apprehension of those which are to come. This is nothing more or less than foresight. So foresight tends to prudence and skill, and they exercise both, and with reference to the future. The goldfinch does not build her nest in the hole of the tree, or in the crotch of the limb; but *hangs it* with *exquisite skill* on the slender *waving, outward branch*, where no animal, or larger bird, or any depredator, can be sustained. She is not more timid than others; why does she invariably thus build? What makes her *"impulses"* differ from those of other birds, and always in the *same manner*?

Jenner, too, has grouped, in admirably descriptive language, many of the peculiarities exhibited by animals and birds before approaching storms, some of which exhibit foresight, and others not.

Perhaps the rooster, who keeps ceaseless watch over his harem, is the most reliable weather-watcher we have. In my earlier days, when it was the practice to keep valuable birds of the kind much longer than it now is, and they had opportunity to become *experienced*, it was interesting to observe how closely they watched the weather. I well remember a venerable chanticleer, who, perched on the tree among his hens, would always foretell the coming storm of the morrow, by sounding forth *in the evening*, and *often*, his defiant note. Such note in the evening was invariable evidence of foul weather. And during the night, their earlier and more frequent crowing is often indicative of it. It is, however, in the earlier part of the day, in doubtful cases, that no inconsiderable reliance may be placed on their sagacity. Often, when a storm is gathering in the forenoon, they will announce it by an almost incessant crowing. The habits of an *experienced*, old-fashioned bird, of this kind, will well repay attention; but I can not answer for the Shanghai and other *fancy breeds*.

Jenner says:

> "The leech disturbed, is newly risen
> Quite to the summit of his prison."

Few have had, or will have, opportunities to observe this, but it is strikingly true. It is difficult to conceive how mere condensation, from an increase of vapor in the atmosphere, should be foreseen by the leech in his watery prison. It is obvious, I think, there is an electric change which reaches him, as it does the whole animal creation, the once broken bones, and the joints of Aunt Betty. Thus much of the philosophy of signs.

The barometer is a useful instrument, in connection with observations of the other phenomena. It is especially useful to the sailor, as its indications relative to the winds are much the most certain. But it is not, *alone*, to be relied upon. This is well settled, although the reasons for it have not been understood. Why it should rise sometimes before storms, in opposition to the general rule—or fall at others without rain—or rise occasionally during the heaviest gales, has been a mystery, and impaired the confidence in its accuracy and usefulness even of the class of philosophers of whom Sir George Harvey spoke, in the sentence quoted in the introduction. But, as I have already intimated, it is all very intelligible.

I have said that the barometer has no fair weather standard—the mean of 30 inches at the level of the sea being an *average* of the *fair weather* elevations and the *foul weather* depressions. Its fair weather position, it would seem, must be above the mean, therefore, and as much above as its foul weather depressions are below. But this is not precisely true. Its extreme fair weather range is 31 inches, and it rarely reaches that; while its lowest storm range is down to 28, and is the most often reached of the two. My barometer stands about 40 feet above ordinary high-water mark. It is not a "wheel," but an open, "scale" barometer, and a perfectly good one. Its most reliable fair weather standard is about $30^{30}/_{100}$ inches. It is its *most common summer, set fair position*, but that position is often at other and different elevations, at other periods of the year, during fair weather. The reader must observe for his own locality, and satisfy himself what the most common set fair position for the barometer is, at the different periods of the year, where he resides. When he has ascertained this, he may apply the following principles to illustrate its exceptional action, and in judging of the future of the weather:

1st. *As to its rise before storms.*—Supposing it to have been stationary, at or about a set fair position, *for the period,* and for one or two or more days, a very *gradual* and *moderate* rise is an indication of continued fair weather; and a *sudden* and *considerable rise* is indicative of a storm. If the sudden and considerable rise occurs in the latter part of spring, summer, or early autumn, it indicates a storm of the *first* or *third classes* described in Chapter X., if in winter, a storm of the *first class* only. If the elevation is *very* sudden and considerable, the storm will probably be *severe.* The philosophy of this, according to my present apprehension of it, is, that these storms present an *extended easterly front*—*settle very near the earth*—and *have a rapid progress*—thus accumulating the atmosphere somewhat, in advance of them.

2d. *As to its fall before storms without previous rise.*—This is always very regular before the second class of storms, or polar belts of showers and storms. It is very fairly exemplified in the table from Reid, on page 329. The barometer, so far as I have opportunity to observe, does not rise from a stationary position on the approach of this class of storms. At the commencement of heated, summer, dry terms, my barometer has most frequently ranged at about 30.30, and gradually, but slowly, fallen below 30 inches before the belt of showers arrived, and the term closed. The fourth rule of Dalton (Meteorology, page 183) indicates a similar law in England. It is as follows:

> "In summer, after a long continuance of fair weather, with the barometer high, it generally falls gradually, and for one, two, or more days, before there is much appearance of rain. If the fall be sudden and great for the season, it will probably be followed by thunder."

3d. *It falls frequently and considerably without rain.*—This is owing to the fact that *all* regular, periodic efforts at condensation do not result in rain. The second, third, and fourth classes of storms described, may not (as we have said) *be sufficiently active to precipitate,* although the *series of phenomena* (including the fall of the barometer) may be, in other respects, perfect. Such an instance may be found in Reid's table, on page 329, and on the 11th of the month. But the fall in such cases is not as great, unless the wind be violent.

4th. *It rises during considerable gales.*—But these are of the kind so often alluded to—viz., the N. W., in the northern hemisphere, and the S. W., in the southern; and the *philosophy* of it has been explained, and is observable.

With these explanations, the reader will be able to understand, and practically apply, the barometric changes, in connection with the other phenomena, in forming an opinion of the weather.

The thermometer is also an auxiliary. It *rises,* during the winter half of the year, in the *advance portion of the storm,* and falls when it passes off again; and the reverse is true, as we have seen, when its range is very high in summer. It is, therefore, to some extent, a useful auxiliary, although of minor importance.

The hygrometer is of less importance still. It is not in general use as a practical guide to the changes of the weather, and does not deserve to be.

A question, which has been much mooted, deserves a passing notice in this connection—viz., whether our climate has gradually become ameliorated and milder on the eastern part of our continent, since its settlement. I have not space left for its discussion. Humboldt (Aspects of Nature, page 103) is of opinion that there has been no material change. He says:

> "The statements so frequently advanced, although unsupported by measurements, that since the first European settlements in New England, Pennsylvania, and Virginia, the destruction of many forests on both sides of the Alleghanys, has rendered the climate more equable—making the winters milder and the summers cooler—are now generally discredited. No series of thermometric observations worthy of confidence extend further back, in the United States, than seventy-eight years. We find, from the Philadelphia observations, that from 1771 to 1824, the mean annual heat has hardly risen 2°.7 Fahrenheit—an increase that may fairly be ascribed to the extension of the town, its greater population, and to the numerous steam-engines. This annual increase of temperature may also be owing to accident, for in the same period I find that there was an increase of the mean winter temperature of 2° Fahrenheit; but, with this exception, the seasons had all become somewhat warmer. Thirty-three years' observation, at Salem, in Massachusetts, show scarcely any difference, the mean of each one oscillating within 1° of Fahrenheit, about the mean of the whole number; and the winters of Salem, instead of having been rendered more mild, as conjectured, from the eradication of the forests, have become colder, by 4° Fahrenheit, during the last thirty-three years."

The facts hereinbefore stated show that there is nothing like a *regular* amelioration; that the seasons differ during the same decade, and different decades. The cold decade, from 1811 to 1820, has not been reproduced. But it may be, and we know not how soon. Since that period there has certainly been a change—for even the cold period from 1835 to 1840 did not equal that from 1815 to 1820, nor indeed those of 1775 to 1780 or 1795 to 1800. But as these variations, so far as we are enabled to judge, depend upon the varying influence of the sun's rays, and of volcanic action, it is impossible to say that equally cold periods will not return, during the latter half of this century.

If the influence of the sun was constant, and volcanic action regular, two causes would tend to modify the seasons:

1st. The exposure of the surface to a more effective action of the solar rays, by a removal of the forests, and by drainage. That such action would be more effective upon a surface thus uncovered and drained, can not be doubted.

2d. *The movement of the area of magnetic intensity, and the magnetic pole, to the west.*—There is such a movement, and its progress can be measured by the increase of declination on the east of it, and its decrease on the west. And the effect of it on climate is unquestionable. In all probability it has had an influence upon ours; and a removal of that area and pole still further west—60° or 80°—would change the location of the concentrated trade, and the Gulf Stream, and restore to Greenland the fertility she once had, and which the Faroe Islands now enjoy. And, on the other hand, its removal as far east of its present position would again depopulate Greenland, and render it again inaccessible. But I can not pursue this subject.

Finally, assistance may be derived from the occasional, although imperfect, accounts of the state of the weather elsewhere, which the newspapers afford. I have been much indebted to the Associated Press of New York for intelligence contained in their telegraphic reports. Occasionally they have been very full and instructive.

On this point, however, there is less of reality in the present than of hope in the future. The time must come when the collection and dissemination of meteorological truth, will be deemed an object of national importance, and national duty. Population is increasing, by immigration and propagation, in a rapidly progressive ratio. There has been great danger that it would outrun agricultural production. A short crop this year would have been disastrous to our prosperity—and the danger was imminent. Every description of business, and every financial circle, felt that fever of anxiety it was so well calculated to induce. The importance of extended agricultural production, and the dependence of all classes upon its success, are now in a greater measure appreciated; and none can fail to see the value of a correct understanding of the weather to the agriculturist, how short-sighted soever they may be, in relation to its direct influence upon their own prosperity and happiness.

Our country is, physically, a most favored one. The facts disclosed or alluded to in this volume show that it is without a parallel on the face of the globe; and our facilities for meteorological observation, and the ascertainment and practical application of meteorological truth, are equally pre-eminent. The great extent and unbroken surface of the eastern portion of the continent; its excessive supply of magnetism and atmospheric currents, and the consequent marked character of the phenomena; the existence and prospective increase of telegraph lines over most of its surface; the homogeneous and energetic character of a population united, upon so large a surface, under one government; the freedom of that government from debt, and the excess of its revenue; the possession of a National Observatory, with a competent philosopher at its head; and a national institution, liberally endowed, and adapted to the collection and diffusion of practical and scientific intelligence, give us an opportunity and a capacity for connected observation and investigation, and an ability to profit by it, that no other nation can boast.

We have, too, a just national pride. Our exploring ships have penetrated and made discoveries in both hemispheres, and our travelers have visited successfully every clime; and thus our national interests, and obligations, and pride, demand an organization, practical and permanent, in relation to this subject, and the time will come when we shall have it.

When that time comes—when the present *limited horizon* of each of us is *practi-*

cally extended over the entire country—and when the actual state of the weather over every part of it is known, at the same time, to the inhabitants of every other, and every where *read in the light of a correct philosophy*, prognostication will be comparatively simple and certain; and A PROGRESS will have been made, productive of an amount of pecuniary, intellectual, and social benefit to the people, which can not be overestimated. May it come before the shadows of the night of death have gathered around us, that we may have a more perfect view of that atmospheric machinery which distinguishes our planet from others, and is, with such infinite wisdom, adapted to make it a fit habitation for man!

THE END.

APPENDIX.

Since this work was completed I have received a very valuable publication, entitled, the "Army Meteorological Register." It is a compilation of the observations made by the officers of the medical department of the army, at the military Posts of the United States, from 1843 to 1854 inclusive, prepared under the supervision of the Surgeon-general, and published by direction of the Secretary of War. To this, there is appended a report or general review of the prominent features of American climatology, so far as the basis afforded by the published observation of the army medical Bureau would warrant positive deduction, by Mr. Lorin Blodget, a distinguished meteorologist, accompanied by temperature and rain charts, for each of the four seasons;—exhibiting the various local differences and peculiarities relative to temperature and precipitation in each.

These local differences and peculiarities and contrasts are deduced and delineated by Mr. Blodget with much ability. He was fettered, however, by the prevailing calorific theories, and the unfortunate practice of grouping the phenomena into means for the seasons, Spring, Summer, Autumn, and Winter, which grouping is arbitrary, and comparatively uninstructive. Hence, he failed to discover what the tables and summaries most clearly disclose—the principles and system unfolded in the foregoing work.

But the summaries of this register contain observations made at posts in Western and Southwestern Texas, in Kansas and Nebraska, and in New Mexico and California, where there has been a dearth of such observations hitherto, and enable me to demonstrate, more conclusively, and I think so that none can fail to understand it, the truth of the philosophy I have endeavored to exhibit.

To do this, I will take a *year*,—divide it into two seasons, the periods of northern and southern transit, the only natural and correct division—and note the phenomena in each, as each progresses.

And I will take the year 1854, because that is the last year for which the record of observation is complete; because it had marked peculiarities which are remembered; and because I have alluded to those peculiarities, and those allusions should be confirmed or disproved by the record. Unless I mistake exceedingly, the confirmation will be found signal and convincing.

I have assumed, pp. 187, 351, that the transits were greater in some seasons than others; that the drought of 1854 was owing to an extreme northern transit, or to an extension west of the concentrated counter-trade, or both, leaving us less supplied with moisture than usual.

In point of fact, it appears from these observations that it resulted from *both* causes, operating *connectedly*; and the annals of Science rarely furnish a more striking instance of analogical inference proved true by subsequent investigation.

Commencing then with the commencement of the northern transit about the 1st of February, we are enabled to trace the then location of our concentrated trade, and its subsequent progress to the north till August, and its influence upon temperature and precipitation. And we can also trace the situation during the same period, of the intervening drought, and the inter-tropical belt of rains, and the extension of the latter north over Florida and the cotton-planting States.

On the 1st of February, 1854, our counter-trade was somewhat more concentrated on its extreme winter curve, over the Southern States, than usual. Its line of excess reached up from Fort Brooke, on the peninsula of Florida, to the north-west, a little east of Pensacola on the gulf, cutting Mount Vernon Arsenal north of Pensacola, and extending thence north-westwardly on to Eastern Louisiana, and curving thence and passing N. E. or E. N. E., to the Atlantic, about the waters of the Chesapeake Bay. It thinned out to the west over New Orleans and Baton Rouge, supplying them moderately, but did not extend to the forts of Texas on the west, nor the posts in the Indian Territory at the N. W. It was east of Fort Towson, which is the south-eastern one. It did not reach St. Louis on the north, nor extend north of the Ohio River, as will appear from the tables hereinafter given. The following cut shows substantially its situation on the 1st of February.

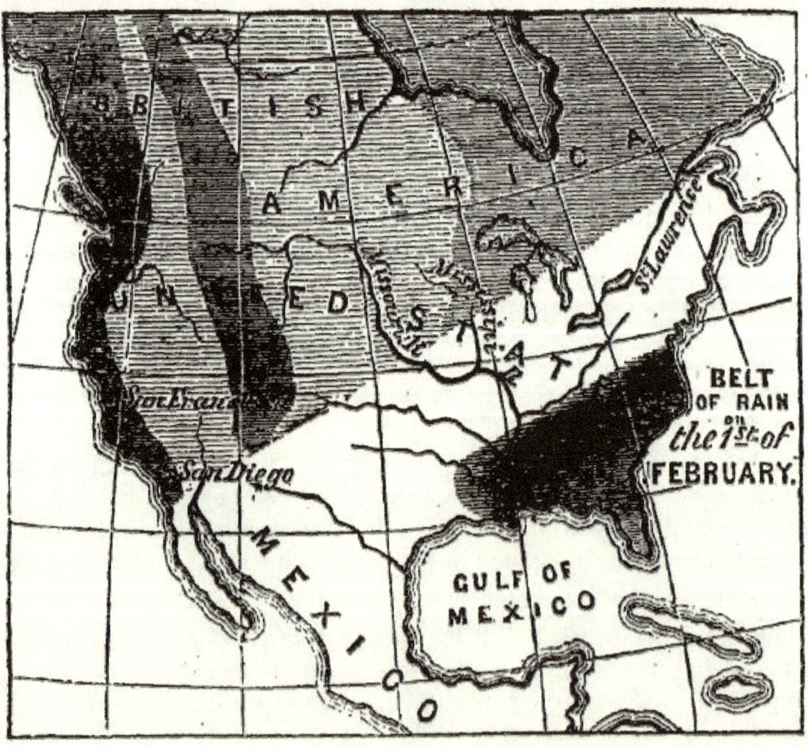

Now, during the month of January, we find the following state of things. *Under* this concentrated trade, the temperature was above the mean, even if Forts Monroe and McHenry on the Atlantic are included; but Mr. Blodget discredits their returns, and some others which do not conform to general results. On the west and north of its curving line, both precipitation and temperature were below the mean.

Under the counter trade, we have the following stations, with their actual and mean temperature. I have inserted the temperature for several subsequent months, to show a depression in April.

TABLE I.

	LAT.	LON.	JAN.	FEB.	MAR.	APRIL.	MAY.	JUNE.	JULY.
Fort Moultrie	32.45	79.51	**50.83**	53.09	62.72	62.76	73.35	78.55	82.06
Mean of 28 yrs.			50.36	52.41	58.68	65.44	73.42	79.01	81.72
Fort Pierce	27.30	80.20	**67.91**	67.33	73.01	71.10	78.41	82.09	84.16
Mean of 5 yrs.			62.75	64.42	69.77	73.63	76.92	79.02	82.50
Fort Meade	28.01	82.00	**63.75**	63.33	70.64	68.10	76.31	79.10	80.17
Mean of 3 yrs.			58.40	63.23	69.02	69.89	76.69	78.24	79.76
Fort Brooke	28.00	82.28	**62.94**	62.36	70.06	70.07	77.49	80.51	81.08
Mean of 25 yrs.			61.53	63.54	67.72	71.82	76.64	79.46	80.72
Fort Myers	26.38	82.00	**67.56**	67.39	73.74	71.07	79.13	82.35	81.91
Mean of 4 yrs.			63.39	67.98	72.19	73.86	80.13	81.25	82.87
Key West	24.32	81.48	**71.75**	71.95	76.56	73.89	80.84	83.34	83.30
Mean of 14 yrs.			66.68	68.88	72.88	75.38	79.10	81.63	83.00
Fort Barrancas	30.18	87.27	**54.71**	54.56	64.98	62.93	75.40	81.00	84.55
Mean of 17 yrs.			53.61	55.58	61.80	68.51	75.45	80.80	82.26
Mt. Vernon Ars'l	31.12	88.02	**51.52**	53.18	65.24	62.30	74.64	79.17	78.90
Mean of 14 yrs.			50.44	53.69	60.26	66.87	73.92	78.03	78.62
Baton Rouge	30.26	91.18	53.43	56.48	66.24	64.63	75.10	80.61	80.09
Mean of 24 yrs.			53.47	55.02	61.93	69.30	75.60	80.56	81.81

It will be seen that the temperature was above the mean in January at every post except Baton Rouge, and there it was at the mean. We shall see hereafter that Baton Rouge was near its western line.

Under this trade during this month, and at the same posts, the fall of rain was as follows, compared with the mean:—

TABLE II.

	JANUARY.		FEBR'Y.		MARCH.		APRIL.		MAY.		JUN.	JUL.
	1854.	Mean.	1854.	Mean.	1854.	Mean.	1854.	Mean.	1854.	Mean.		

Key West	1.77	2.86	2.55	1.38	0.51	4.21	2.99	1.55	3.14	2.58	4.54	3.45
Fort Myers	1.15	3.90	4.70	2.16	0.20	4.60	2.75	3.14	5.65	3.33	6.75	9.70
Fort Brooke	3.88	2.20	6.89	3.01	2.44	3.37	8.82	1.95	6.21	3.24	9.44	15.53
Fort Mead	1.30	1.07	2.21	1.01	1.85	1.64	3.19	1.78	10.51	5.34	7.24	8.55
Fort Pierce	3.55	4.45	3.40	2.72	1.05	3.01	7.00	3.85	5.70	4.27	6.63	4.97
Fort Barrancas	3.45	3.87	5.55	4.95	7.21	5.87	0.50	2.94	3.47	4.05	3.39	5.43
Mt. Vernon Ars'l	11.01	6.80	12.83	6.04	6.22	4.59	1.96	4.21	4.45	4.62	6.72	6.13
Baton Rouge	2.85	5.26	5.50	4.91	6.15	4.68	3.58	5.22	8.05	5.18	4.00	6.55
Fort Moultrie	3.80	2.39	2.84	2.33	0.25	4.06	2.20	1.75	3.70	4.08	4.20	5.69

It will be observed that in February the counter-trade and extra-tropical belt had moved up from Key West, and a drought, which sometimes intervenes between the concentrated counter-trade and the inter-tropical belt, appeared there in February and March. In April, the inter-tropical belt appeared at that point, and went on increasing till September. As the counter-trade commenced moving north in February, an increased precipitation above the mean commenced at all the more southern stations under the concentrated-trade—an earnest of that irregularity which followed, and marked the season as the most excessive of the century.

In March, the intervening drought appeared at the other posts on the peninsula, and also at Fort Moultrie, followed *much more closely than usual*, by the inter-tropical belt of rains. In April, the drought appeared at Fort Barrancas and Mount Vernon Arsenal (the wave of precipitation having moved to the west), and slightly in comparison at Baton Rouge.

If now we look at the condition of things, *west* and *north* of the curving line of concentrated trade, from Fort Brown, at the mouth of the Rio Grande, in South-western Texas, through that State, the Indian Territory, Arkansas, Missouri, Kentucky, and Northern Pennsylvania, to the Atlantic, we find the thermometer every where in January below the mean. The following table will show this, and the precipitation for that month and February:—

TABLE III.

	JANUARY.		FEBRUARY.		MARCH.		Rain in January.	Rain in February.
	1854.	Mean.	1854.	Mean.	1854.	Mean.		
Western Texas								
Fort Brown	59.34	60.41	62.45	63.63	71.87	68.95	0.45	1.50
Fort Ewell	50.47	52.92	58.12	57.61	70.34	67.00	0.22	2.86
Fort Inge	47.24	49.46	56.04	55.39	67.54	62.63	0.20	2.15
Indian Territory								

Fort Towson Forts Gibson, Washita, and Ar-buckle, in much the same propor-tions.	36.32	43.14	49.29	45.97	59.55	53.40	1.01	2.00
Arkansas								
Fort Smith	33.92	40.18	47.01	43.89	57.01	51.58	1.37	2.05
Missouri								
St. Louis Arsenal	25.47	31.44	36.66	33.43	46.10	42.30	0.65	2.40
Kentucky								
Newport Bar-racks	31.75	34.04	39.60	36.94	46.74	45.46	3.20	5.30
Pennsylvania								
Allegheny Arsenal	29.08	29.25	33.49	31.16	40.36	39.02	2.23	2.33
Delaware								
Fort Delaware	32.38	33.67	34.56	35.84	43.18	42.90	2.30	5.45
New York Harbor								
Fort Columbus	28.71	30.18	28.17	30.44	36.17	38.28	2.60	4.00

We find, also, from this and table first, that every where, except at Fort Brown, and upon the Atlantic coast, the temperature had risen above the mean in February.

The situation of the belt which supplied the western coast in winter, and its excess of precipitation, are also represented upon the cut. The intervening area was not without counter-trade and precipitation—the latter, of course, greatest over the area of intensity—but they were *comparatively* less, as the tables will show.

The following cut and table show the situation of the concentrated counter-trade in March.

TABLE IV.

	JAN.	FEBR.	MAR.	APRIL.	MAY.	JUNE.	JULY.
Fort Barrancas, Pensacola Bay	3.45	5.55	**7.21**	0.50	3.47	3.39	5.43
Mean	3.87	4.95	5.87	2.94	4.05	4.66	6.80
Baton Rouge, Louisiana	2.85	5.50	**6.15**	3.58	8.05	4.00	6.55
Mean	5.26	4.91	4.68	5.22	5.18	5.52	7.42
Fort Towson, Indian Territory	1.01	2.00	**5.10**	2.22	Recr'd	stops	here.
Mean	3.13	2.97	4.38	5.33			
Fort Gibson, Indian Territory	0.30	1.43	**7.83**	3.16	7.67	2.80	0.21
Mean	1.33	2.26	2.54	4.19	4.65	4.30	2.75

Fort Smith, Arkansas	1.37	2.05	**7.05**	6.55	6.25	2.26	1.02
Mean	1.96	2.17	2.92	5.10	4.46	4.74	3.82
St. Louis Arsenal	0.65	2.40	**7.10**	4.30	4.65	2.20	1.70
Mean	1.93	3.37	3.82	4.16	4.88	6.94	0.04
Newport Barracks, Kentucky	3.20	5.30	**8.10**	2.10			
(No Mean given.)							

We see from this table that its focus had extended west in Florida over Fort Barrancas, and over Baton Rouge in Louisiana; N. W. to Forts Towson and Gibson in the Indian Territory, and Smith in Arkansas; north to St. Louis Arsenal at St. Louis, and to Newport barracks in Kentucky; but it was spread over a larger surface east of the mountains. Its greatest progress for the month, was a west and north-west progress.

In April, we find it had progressed rapidly west and north-west, and its position is shown by the following cut and table.

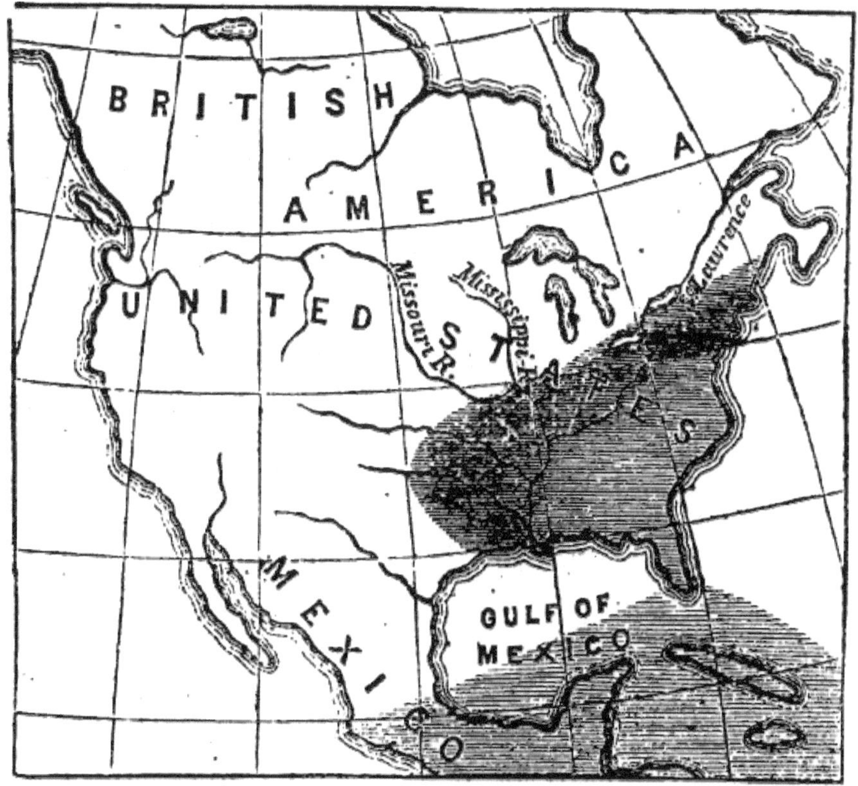

TABLE V.

	JAN.	FEBR.	MAR.	APRIL.	MAY.	JUNE.	JULY.
Fort Riley, Kansas	0.00	0.94	1.86	**4.55**	4.35	1.10	0.00
Fort Leavenworth, Kansas	0.04	1.78	1.33	**3.35**	**5.55**	4.50	0.18
Mean	0.72	1.01	1.61	2.74	3.62	5.80	3.15
Alleghany Arsenal, Pittsburgh	2.23	2.33	2.82	**4.21**	2.24	2.06	1.45
Mean	2.18	2.17	2.70	3.10	3.58	3.56	2.97
Fort Columbus, New York Harbor	2.60	4.00	0.70	**8.80**	7.70	2.20	1.90
Mean	2.78	2.92	3.44	3.33	4.78	3.46	3.17
Fort Independence, Boston	2.50	3.36	2.55	**5.40**	4.28	2.00	
West Point.	3.52	5.04	2.81	**10.53**	2.00	1.62	
Mean	3.50	3.44	3.71	4.55	6.18	4.79	

We see, too, that both east and west of the mountains, its focus of precipitation was one month in advance of the mean. At all the stations where the greatest fall was in March, it should have been in April, and the fall at those points was greatly in excess of the usual quantity. And the same was true of stations reached in April. The concentrated trade, instead of spreading out, and precipitating over the whole south-eastern portion of the continent (its normal condition), was gathered into a wave of greater volume, resulting in greater precipitation, and was rapidly hastening its curve to the west over Texas, and to the north-west over the Indian Territory, and northward on its usual curve to the north and east of them.

The observations for April disclose another singular and instructive condition. The temperature, that had every where been above the mean in March, fell below it in April under the concentrated trade. And snow fell on three days in some localities, and four in others.

Along the Ohio River, it fell to the depth of 8 to 10 inches on the 17th, and east of the mountains to a greater depth on the 18th, one day later. It fell to the depth of 4 inches at Marietta on the 29th also. Dr. Hilldreth, American Journal of Science for March, 1855, says:—

"It is a singular fact that the deepest snow, 8 inches, fell on the 17th of April, and at the head waters about Pittsburg over a foot. Also, on the 29th of the month, at Marietta, 4 inches, a very rare occurrence." This depression of the temperature was quite general, but the fall of snow was local. The latter was north of a line drawn from Fort Laramie, at the base of the Rocky Mountains, in an E. S. E. direction—north of Forts Kearney and Leavenworth, and of St. Louis, but south of Newport barracks in Kentucky, and from thence to the Atlantic. Snow fell at every station north of this line, at no station south of it. The depression of temperature, however, was experienced over the continent, east of the Rocky Mountains, under, and south of, the belt of precipitation. Now what occasioned this general depression of temperature, and local fall of snow? It will not do to say, as perhaps some calorific theorist may be inclined to say, because the concentrated trade had been carried up where it was cold, a month too soon; or that the sun had heated the land in advance of it, and drawn it up.

For, 1st, it might be asked how, if it was warm enough to draw it up, could it be cold enough to make it snow; or, 2d, how happened it to start, when, as we have seen, it was warmer than the mean under it, and colder than the mean to the north and west of it, when it commenced its journey?

But again, it snowed at posts north of the line, while the thermometer remained above the mean; and the thermometer fell below the mean down to Fort Brown in south-western Texas, and at Key West in the southern part of Florida; and what is more remarkable still, at Key West, Fort Barrancas, and every other south-eastern station, except Forts Brooke and Moultrie, it not only fell below the *mean* of the month, but *below the actual temperature of March*. (See Table I.) At Forts Brooke and Moultrie it did not rise above that temperature. West of the Rocky Mountains the depression was not felt; nor at stations north, or north-west of the belt of precipitation.

It is obvious, the calorific theory can furnish no rational explanation of this matter; for the reason that, whatever the cause, it operated not only under, but south, and far south of the belt of precipitation. It could not have been spots upon the sun, or other general cause, for then it would have operated in New Mexico and California, and at the north-western stations. It operated most intensely in Florida and the South-Eastern States, which approach most nearly the volcanic areas of South America and the West Indies. I believe it to have been occasioned by volcanic action affecting the local magnetism of our intense area; but it is a most important development, and should be thoroughly investigated. We may find in it the key to the mysterious, but unquestionable, influence of volcanic upon magnetic action; and I hope the distinguished surgeon-general will cause the records of that month to be published "in extenso."

In May and June, the trade became more concentrated, a perfectly developed belt from the Rio Grande to the Lakes and British possessions, and doubtless to the Atlantic, with every where a central focus of excessive precipitation, gathering to itself in one vast wave the current that should have been spread out over the whole country; and leaving every where on its eastern and southern borders, down to the northern edge of the inter-tropical belt of rains—(which extended up to lines drawn from Baton Rouge to Charleston)—a *perfectly well developed* and *defined drought*. That drought will long be remembered. The following cuts show, approximately, the location of the belt of precipitation and drought for those months, and the table which follows will show their correctness.

The tables also show that this wave was occasionally a double, or divided one—evinced by an intervening *partial* precipitation. Tables IV., V., and VI., also show the commencement of the drought at the several stations, as the wave moved to the west and north.

MAY.

BRITISH
AMERICA
UNITED
Missouri
Miss.
STATES
AREA OF DROUGHT
MEXICO
GULF OF MEXICO

BRITISH
AMERICA
UNITED
STATES
JUNE
MISSOURI
AREA of DROUGHT
MEXICO
GULF OF MEXICO

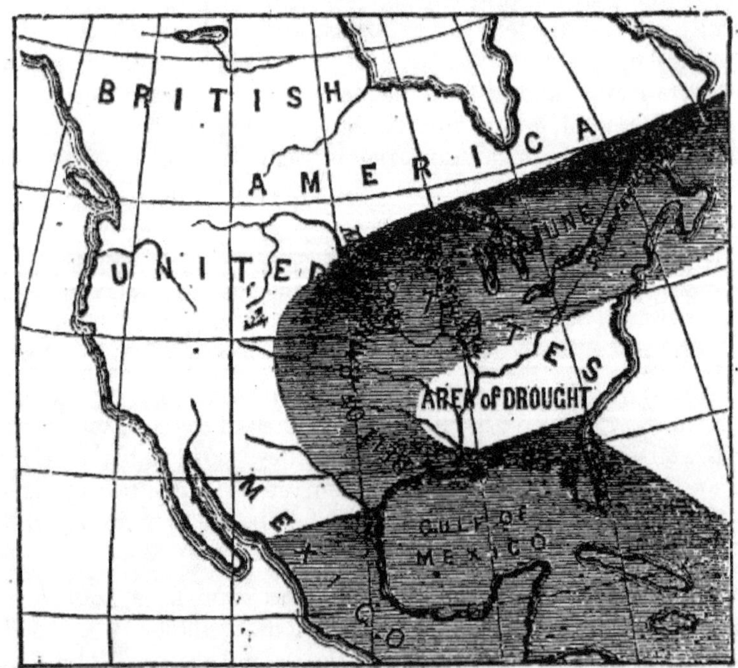

TABLE VI.

	JAN.	FEBR.	MAR.	APRIL.	MAY.	JUNE.	JULY.	AUG.	SEPT.
Fort Brown	0.45	1.50	1.15	0.05	4.10	**7.65**	4.25	5.00	11.31
Mean	1.61	2.25	1.20	0.56	2.21	4.55	1.95	2.76	6.73
Ringgold Barracks	0.70	1.69	0.22	0.00	2.83	**10.98**	4.06	1.58	3.02
Mean	1.24	1.18	0.72	1.08	2.09	3.47	3.18	1.50	3.22
Fort Merrill	0.11	1.99	0.05	1.16	**7.66**	4.70	5.44	3.13	5.01
Mean	0.23	2.09	0.09	1.62	3.43	4.10	6.13	3.40	4.60
Fort Duncan	0.05	0.69	1.50	0.00	2.53	**6.83**	0.83	0.90	4.81
Mean	0.26	1.27	1.34	0.71	1.50	5.63	3.35	0.93	3.28
Fort Inge	0.20	2.15	3.00	0.75	**3.88**	2.09	0.97	1.67	4.80
Mean	0.64	2.21	1.79	1.26	3.01	5.38	3.66	2.02	2.21
Fort McKavet	0.01	0.77	2.10	0.28	**3.72**	0.15	2.91	0.04	3.86
Fort Belknap	0.11	1.10	1.42	1.75	4.97	**8.33**	0.00	0.75	1.53
Fort Massachusetts, Northern New Mexico					**3.93**	0.24	2.14	2.61	1.53
Fort Kearney	0.23	1.33	1.87	2.56	4.15	**5.40**	3.51	1.18	4.60
Mean	0.50	0.48	1.55	2.68	6.57	4.36	5.07	2.62	1.83
Fort Laramie	0.18	0.40	0.80	3.98	**4.46**	3.67	3.26	1.27	1.60
Mean	0.27	0.71	1.37	1.93	5.39	2.95	1.83	0.92	1.33
Fort Ridgley	1.20	0.01	1.18	2.83	**6.84**	2.70	2.49	2.28	2.58
Fort Snelling	0.72	0.03	1.03	2.51	**4.30**	3.31	3.92	1.75	6.55
Mean	0.73	0.52	1.30	2.14	3.17	3.63	4.11	3.18	3.32
Fort Ripley	0.67	0.03	0.79	0.97	**4.34**	3.68	0.62	1.69	4.40
Mean	0.86	0.37	1.80	1.42	3.09	5.15	5.20	2.27	4.92
Fort Mackinac	2.59	1.23	1.56	1.04	2.65	**6.35**	5.67	4.26	3.22
Mean	1.25	0.82	1.14	1.21	2.32	2.81	3.20	2.87	2.97
Fort Brady	2.49	1.18	1.34	2.14	**3.61**	1.23	3.21	3.86	3.18
Mean	1.84	1.13	1.37	1.83	2.24	2.83	3.73	3.39	4.33
Fort Niagara	1.63	2.52	1.87	2.25	3.90	1.71	4.08	1.52	2.61
Mean	2.25	1.89	2.12	2.20	2.55	3.28	3.49	3.04	3.95

But the belt of trade continued its progress to the west and north, and during the months of July and August the drought extended in both directions, reaching, in August, from Mississippi, Alabama, Georgia, and South Carolina, to the Lakes, and from the Rocky Mountains to the Atlantic. Its position is shown by the following cut, and the position of the belt of precipitation by the following table.

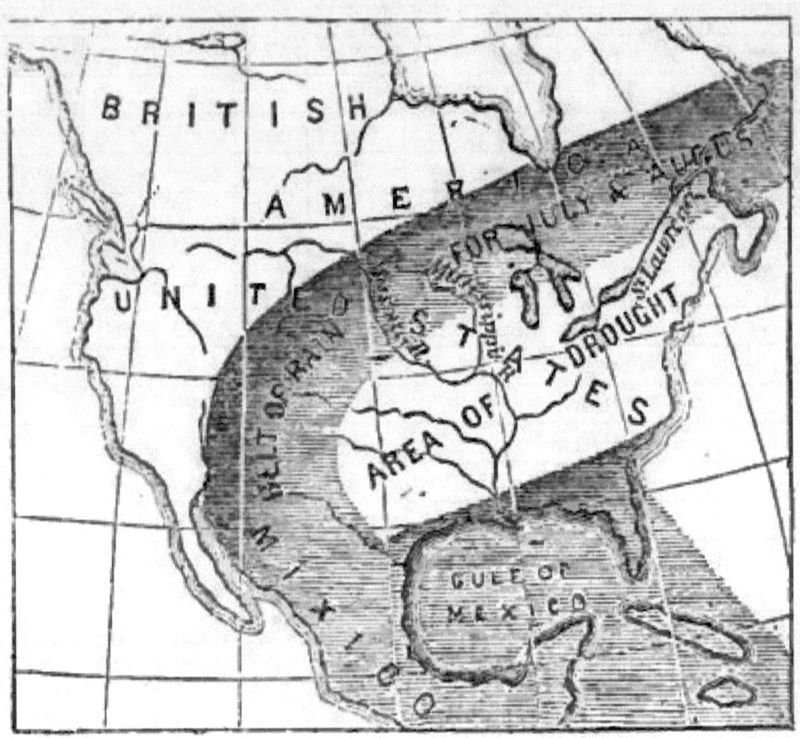

TABLE VII.

Situation of the focus of Precipitation in July and August.

	JUNE.	JULY.	AUG.	SEPT.	OCT.
New Mexico.					
Fort Thorne	0.08	2.23	6.01	3.50	0.00
Albuquerque	0.28	2.50	1.19	2.67	1.37
Santa Fe	0.32	4.11	3.86	4.06	2.50
Fort Defiance	1.24	3.94	5.24	3.47	0.62
Fort Yuma	0.00	0.01	2.37	0.17	0.30
San Diego	0.02	0.07	1.35	0.13	0.01
Fort Snelling, Minnesota	3.31	3.92	1.75	6.35	1.23
Fort Brady	1.23	3.21	3.86	3.18	3.40
Fort Mackinac	6.35	5.67	4.26	3.22	2.28

I have not space for all the comment which this exposition is calculated to induce. The reader will not only find in it an explanation of the extraordinary character of the summer of 1854, but will see from the *means*, that it was but an *exces-*

sive development of an ANNUAL PHENOMENON,—THE PROGRESS OF A CONCENTRATED COUNTER-TRADE.

It is not necessary to follow with particularity the return transit. It required no great degree of sagacity to predict, at the time, that the drought would continue in the vicinity of New York till about the 10th of September. The return of the belt to that latitude, was not to be expected before that time, and the drought continued, in fact, until the 9th of September.

Its return progress was slow, and it was every where behind time. The autumn was warm, and so, indeed, were December and January, west of the area of magnetic intensity, although upon, and east of it, there was a depression in December. The retreating but lingering edge of counter-trade, with its excess of snow for the season, caught the Iron Horse, with its train and passengers, upon the prairies of the west, and laid its embargoing hands upon them. Few, if any, can have forgotten the thrilling accounts which reached us from that section, of the sufferings endured by those who were thus embargoed for days and nights, far from the comfortable habitations of their fellow men.

But the return transit, though slow, was extreme, and February and March were exceedingly cold for the season. The transit to the north, again, did not commence as early as usual, and the spring was backward, and the summer cool. Both were without irregularity, and the season was productive. The following table exhibits the temperature on a line of posts, running north and south at the west, during the winter months of 1855, and will illustrate what has been said.

TABLE VIII.

1855.	JANUARY.	FEBRUARY.	MARCH.	APRIL.
Key West	67.18	65.94	70.28	75.09
Mean	66.58	68.88	72.88	75.38
Fort Snelling	17.09	12.62	25.30	49.86
Mean	13.76	17.57	31.41	46.34
Fort Kearney	23.55	25.69	32.86	54.39
Mean	21.14	26.11	34.50	47.13
Fort Laramie	35.85	29.01	36.41	52.94
Mean	31.03	32.60	36.81	47.60
Fort Arbuckle	41.94	39.86	49.09	67.43
Mean	39.10	43.69	53.22	61.85
Fort Belknap	45.92	44.49	53.09	70.00
Mean	42.80	47.47	56.90	65.79
Fort Chadbourne	48.89	45.87	56.68	68.51
Mean	44.29	46.75	58.01	65.52
Fort McKavitt	46.74	44.51	53.66	67.05
Mean	44.75	46.87	57.39	66.25

Fort Merrill	54.51	54.65	61.82	74.50
Mean	54.82	57.20	68.66	73.27
Fort Brown	60.23	61.60	66.24	74.98
Mean	60.41	63.63	68.95	75.05
Fort Inge	52.21	50.63	61.22	74.48
Mean	49.46	55.39	62.63	68.02

The return transit to the south for this winter, 1855-6, has been an extreme one. It is too early yet (Feb. 18th) to write its history, but the extreme southern transit is as obvious as the unusual severity of the cold. The rains which usually fall upon the Southern States are precipitated further south upon the West Indies, and threaten a deterioration of their sugar crop. The snow, and cold winds, and ice, of the middle latitudes, are felt even in Florida. Our sheet of counter-trade has been exceedingly thin, and the barometer has ranged, in fair weather, much below the mean. Occasional, and for a part of the time, *weekly* periods of an increase of its volume, with a corresponding elevation of the barometer, and a consequent moderation of the intense cold, and a storm, have occurred. But those periods have been few and brief. No regular thaw has yet occurred. From the 26th of December to this date, at Norwalk, there have been but two periods when the wind has blown from the south-west with sufficient force to stir the limbs of the trees. There has been no wind from south of that point, or east of north-east; and even our storm-winds, with one exception, have been north of north-east—owing to the situation of the focus of precipitation far to the south of us—and there is reason to fear that a cold summer like those of 1816 and 1836 may follow. If this extreme transit is owing to defect in the influence of the sun, from spots, or other causes, such will probably be the result. If from volcanic action at the south, the influence of that action may cease, and a rapid return transit, and an ordinary season, may follow. Believing in the laws of periodicity in relation to the weather and disease, I planted an early kind of corn (the Dutton), in 1836, and had a crop when few around me succeeded. We must watch this return transit, with hope, indeed, but not without fear, and be wise in time.

There is a mass of other evidence in these summaries which shows the truth of what I have written. There is not a deduction of Mr. Blodget which it will not explain. The ascent of the summer lines of temperature to the west is explained by the diminution of magnetic intensity. Their descent in winter by the location and attractions of the concentrated trade. The excess of precipitation in Alabama and Mississippi by the succession of summer and winter belts. That of the interior of the Atlantic slope in summer, by the showers which fall upon the elevations; and of the coast, by the easterly storms and their attraction of the surface atmosphere of the ocean, at other seasons. But I cannot further particularize. Even the influence of the spots is clearly demonstrated by the observations at *interior stations*, which were unaffected by contiguous oceans or elevations. At Forts Washita, Gibson, Scott, Smith, and others, the years 1847 and 1848 were below the mean. All that evidence, and those deductions, however, I must pass by for want of space, and take leave of the subject.

Lector House believes that a society develops through a two-fold approach of continuous learning and adaptation, which is derived from the study of classic literary works spread across the historic timeline of literature records. Therefore, we aim at reviving, repairing and redeveloping all those inaccessible or damaged but historically as well as culturally important literature across subjects so that the future generations may have an opportunity to study and learn from past works to embark upon a journey of creating a better future.

This book is a result of an effort made by Lector House towards making a contribution to the preservation and repair of original ancient works which might hold historical significance to the approach of continuous learning across subjects.

HAPPY READING & LEARNING!

LECTOR HOUSE LLP
E-MAIL: lectorpublishing@gmail.com